Darwinism Debunked

Two Classic Apologetics for Creationism in One Volume

DARWINISM DEBUNKED

Two Classic Apologetics for Creationism in One Volume

By Philip Mauro & John D. Charles

TOLE PUBLISHING

Evolution at the Bar first published in 1922; *Fallacies of Evolution* first published in 1917

ISBN:978-1-948696-09-8 (Paperback)

Front cover image: Adobe Stock: Charles Darwin ©nickolae

Printed in the United States of America
TOLE PUBLISHING, PO BOX 1098, MORGANTOWN, KY, 42261-8411
www.tolepublishing.com

Contents

EVOLUTION AT THE BAR
by Philip Mauro

FALLACIES OF EVOLUTION: A SERIES OF THREE LECTURES
by John D. Charles

EVOLUTION AT THE BAR

by Philip Mauro

Forward

The writer of this article was for upwards of twenty years a believer in and student of the philosophy of materialism; but after his conversion to faith in Jesus Christ he rejected it, first because it was found to be contrary to the Scriptures, and second, because, upon careful investigation, it was also found to be opposed to every pertinent fact of history and natural science.

As regards "evidence" in support of the theory, there is none that would be admitted in any court of law. It rests wholly upon unprovable assumptions, and upon highly speculative and far-fetched inferences. The evidence against it is abundant and convincing; and while the subject is so vast and complex that we can present, in an article like this, only a small part of the counterproof, yet we can give all that is needed in order to show that the theory is not only unsupported by proof but is opposed both to the facts of science and to statements of Holy Scripture.

Our design, in writing these pages (originally prepared for use in the compilation of a Bible Cyclopedia by the Bible Union of China) was to set forth the main features of the theory of Evolution in such a way as to make it easy to be understood by the unlearned. The exponents of science and philosophy usually adopt a style and vocabulary which effectually hide their meaning from "the common people," and which are well calculated to produce the impression that the subjects they discuss are too mysterious and profound to be understood by any but the few who (like themselves) are gifted with intellects of a superior order, and possessed of knowledge unattainable by the ordinary man.

But the truth is that--when we disregard mere refinements of detail, and technicalities of a non-essential character--the doctrine of Evolution in general, and that of the Origin of Species (the Darwinian hypothesis) in particular, can be set forth "in words easy to be understood," and can be understood by persons of ordinary intelligence and of common school education. And furthermore, the scientist and philosopher have no facts upon which to base their conclusions except such as are matters of common knowledge or are accessible to all men through textbooks and cyclopedias. We fully concede to experts their special competence in investigating, clarifying, and setting forth the facts; but, in the all-important matter of drawing conclusions from those facts, the expert has no greater ability than the ordinary persons, of whom juries--which in common-law cases are the sole judges of the facts--are composed. It is for the benefit of these that we are now writing; and in summoning Evolution to stand trial at the bar of ordinary common sense, our own function will be mainly to present the pertinent facts as fully and concisely as possible.

As regards the reasons commonly advanced in support of the doctrine of Evolution it is one of the most palpable weaknesses of the case that the alleged "evidences" for it has to be sought in the darkest corners of creation and in the remotest regions of time and space; and further that, when brought into the light of honest inquiry, they cannot be recognized, by ordinary persons, as having any relation at all to the doctrine they are cited to sustain. For Evolution is set forth as a cosmic process---that is, a law operating always and everywhere. It is either that or nothing. But, if so, then the evidences of it would be always and everywhere apparent. Whichever way we might look they would force themselves upon our notice, in countless numbers and endless varieties of forms. The proofs would be so abundant that the demonstrator of the doctrine would never get to the end of them; whereas, as the case actually stands, the efforts and the ingenuity of the evolutionist are mainly occupied in trying to account with plausibility for the total lack of evidence in nature for the support of his doctrine.

Another thing which must impress every fair-minded investigator of this modern theory is the fallacious character of the reasoning often employed by its advocates. Take the case of the now-extinct varieties of a horse having three and four toes. An appeal is often made to the remains of those creatures as if they proved the whole case of Evolution; whereas they do not even prove that the existing varieties of the equine species were derived from those extinct forms. There is nothing whatever to forbid the idea that the present varieties of the species existed at the same time with those now-extinct forms. Proof of connection between them, and of derivation of the one from the other, is wholly lacking. But even if such connection were established, it would not tend in the least to prove the evolution of one species from another, which is the matter in dispute. That many varieties of a common species can be produced is a fact so abundantly in evidence in both vegetable and animal kingdoms as to create a strong presumption that, if it were possible to cross the boundary lines of a species, there would be abundant evidence of that also. But the fact is that, with all nature under observation, and with the plain records of the fossiliferous rocks, not one transitional form to help bridge the gulf between one species and another has ever been found. The four-toed horse is as much a "horse" as the one-toed variety. And Mr. Darwin was himself compelled to concede all that we here point out. He said (Life and Letters, Vol. III. p. 25): "There are two or three million of species on earth--sufficient field, one might think, for observation. But it must be said today that, in spite of all the efforts of trained observers, not one change of a species into another is on record." This statement can be made with even greater confidence now, after a lapse of over half a century since Mr. Darwin made the above admission.

It is vain, therefore, for the evolutionist to think he can ride upon the four-toed horse to a successful demonstration of his theory.

The Wisdom of This World

The doctrine of Evolution is doubtless the culminating effort and fruit of "the wisdom of this world"; and our thought about it is that God will make use of it thereby to exhibit the utter "foolishness" of human wisdom. Never has there been a cosmic philosophy set forth with such pretentiousness or backed with such authority by "Science"; and never has there been a doctrine so audaciously proclaimed in direct and defiant opposition to the truth of Creation, revealed in the Holy Scriptures. Therefore, the theory of Evolution is pre-eminently suited to exemplify the Scripture, "Hath not God made foolish the wisdom of this world?" (1Co.1:20.)

Framingham, Mass.
April 1922.

Chapter I

The Theory Defined

"Evolution" is a philosophical and speculative theory, of recent origin, whereby it is sought to account for the various elements and compounds of the inorganic world, and also for the countless species of living creatures in the organic world.

By the "inorganic world" is meant the elements and compounds, as minerals and gases, which are without life; and by the "organic world" is meant organisms (plants and animals) which have life.

Although sometimes spoken of as a "scientific" theory, Evolution is not scientific; for science has to do only with facts. Evolution belongs wholly in the realm of speculative philosophy.

The basic assumption of this theory is that all things in nature--living and not living--had a common origin; and that all the diverse elements, compounds, and organisms were developed by the cumulative effect of changes, in themselves imperceptibly small, all of which changes were brought about by the energy of "forces resident in nature."

The theory assumes the existence of Matter and Force, without attempting to account for the origin of either. Matter is supposed to have existed originally in a perfectly simple and undifferentiated condition. Its form is supposed to have been that of an exceedingly tenuous, highly heated mist or vapor, filling all space. Force is also assumed to have been exceedingly simple at the first, being nothing more than a tendency on the part of the entire mass of undifferentiated Matter to keep in motion. As to where Matter came from, and Force, and the tremendous uniform Heat, necessary to keep Matter in a gaseous state, the theory is silent.

The theory further assumes that, at some time, and for some unexplained reason, the motion of the particles of matter began to take different directions, and that, by the radiation of the heat of parts of the mass, liquefaction and ultimately solidification resulted. Where the heat so radiated could have gone--seeing that all parts of infinite space were supposedly heated alike--is not explained; nor how, in a perfectly uniform mass, parts could assume a permanently solid form, and other parts a normally liquid form, while other parts remained normally gaseous. Upon the theory of Cosmic Evolution, all gases and liquids should long ago have evolved into solids.

These great changes in Matter are supposed to have been accompanied by equally notable changes in Force. Differences of "Environment" having now arisen, of which differences the theory has no explanation, the effects of Force or Energy would be influenced thereby, in such wise as to produce diversities of forms, until, by the continuous operation of those processes, with ever-increasing ramifications and complexities, the infinite varieties of creatures, animate and inanimate, which now compose the universe, came to be what they are.

Such are the words by which the theory of Evolution is set forth, but the only clear thing about them is that they do not explain the origin of the universe or of any of its parts.

Other principles are called to the aid of Evolution at different stages of the cosmic process; e.g. Heredity, Environment, Natural Selection, Struggle for Existence, Survival of the Fittest, Transmission of Acquired Characters, etc. With these auxiliary factors we have but little concern, our object being to inquire what, if any, foundation in fact there is for the basic theory. If that falls, the auxiliary factors must fall with it.

According to Mr. Herbert Spencer, the leading exponent of the theory, evolutionary changes are of three principal sorts: (1) a change from a less coherent to a more coherent state; (2) a change from a more homogeneous to a less homogeneous state; (3) a change from a less definite to a more definite state.

Le Conte defines Evolution as "(1) continuous progressive change, (2) according to fixed laws, (3) by means of resident forces."

It is important to note the expression "resident forces," which excludes the idea of a Creator acting in or upon the universe.

Such is the theory in its broad outlines; and it is evident that thus far it is wholly imaginative and speculative, every essential feature being assumed without a particle of proof. Indeed it may be clearly seen that the theory is self-contradictory, as in assuming that (under the supposed conditions) latent heat could discharge so as to permit concentration to take place when there were no cooler regions into which it could discharge.

Further, it is self-evident that the action of Infinite Wisdom and Power would be as much needed for the creation of the supposed Matter and Force, with their supposed capacity for development and diversification, as for the creation of separate elements, compounds, and living species. In fact, both Darwin and his co-laborer Wallace had to admit that it was necessary to concede, at various points in the supposed evolution of the world, as well as at the starting point, the working of an outside power, a power not resident in matter. From this admission it follows that there is nothing "unscientific" in the doctrine of Creation by an intelligent Creator.

"Cosmic" and "Organic" Evolution

Nature is seen to exist in two great departments, one comprising things having life, the other things not having life. The former is the "organic" department of nature, the latter the "inorganic." Between these two departments is an impassable gulf. Evolutionists have to concede this; for as Mr. Huxley said, "The present state of knowledge furnishes us with no link between the living and the not-living."

This is a fatal admission; for assuredly, if the entire organic kingdom emerged out of the inorganic, there would be innumerable "links" between the two. It is simply impossible that all traces of such a stupendous transformation should have been obliterated.

To accommodate the theory to this state of the division of nature, Evolution has been correspondingly divided into "Organic Evolution" and "Inorganic" or "Cosmic Evolution." Thus, we have, at present, two distinct Evolutions, each rigidly confined to its own department of nature. The original Evolution, which evolved living creatures out of inanimate matter, no longer exists. It has gone entirely out of business and has ceased to exist from the time, whenever it was, that the world of living creatures was separated, by an impassable barrier, from the not-living. It would follow that Evolution is not what it once was. Having once crossed the line which separates the living from the non-living, it has lost the power to do so again.

Cosmic Evolution

Cosmic Evolution, or Evolution as it is supposed to operate in the universe at large--the starry heavens, the earth and sea, and air--calls for but brief notice in this article. Proof of the existence, either now or in past ages, of any such "law" as that of Evolution, is altogether lacking. Suffice it, therefore, to say that if, anywhere in the universe, at any stage of its existence, undifferentiated matter has been gradually transformed by means of resident forces, into the various substances of earth, sea, and sky, with their widely different and often antagonistic properties, there has never been discovered by mortal man the shadow of a shade of a proof thereof. The results of all investigations that have been made up to the present hour bear accordant witness to the fact that stability of forms and of the properties of inorganic substances, is the fixed rule of nature. Those who accept the idea of Cosmic Evolution must need to do so without any evidence whatever to support it, for none exists.

Organic Evolution

How then stands the case with respect to Evolution "Organic Evolution"? Is it any better supported than "Cosmic Evolution"? In this field, it will be necessary to make a closer examination of facts and phenomena; for living creatures do undergo changes. In fact their existence is one of continuous change.

What characterizes the organic department of nature is the existence of individuals, each living an independent life of its own, and each having its own life-history. Each of these individual organisms comes suddenly into being; it goes through various stages of growth until maturity is reached; it reproduces its kind; it declines and suddenly ceases to exist. This is what we find throughout the entire organic field. But there is nothing in the inorganic department of nature which even remotely resembles this life-story of individuals. That field will be searched in vain for anything out of which the details of the organic world, comprising several millions of species, each with an infinitude of structural and other peculiarities, could conceivably have been evolved. Yet, the theory of Evolution, as a universal or cosmic process, requires us to believe that the entire organic world emerged, at some past era, from the inorganic. Surely, if such were indeed the case, then the latter would contain abundant evidence thereof, showing how individual entities, with their characteristic life-changes, came into existence. And not only so, but we should also find everywhere inorganic groupings of atoms gradually reaching forth towards organic existence, and most certainly it would be possible by laboratory methods to transform the one into the other.

Due notice should also be taken of the striking fact that the beginning of the existence of each living creature is sudden, that its term of life is short, and that its changes are rapid. Whereas Evolution requires a very gradual coming into existence, exceedingly long histories, and changes of prodigious slowness. The fact then is that, in the field of the living, as in that of the not-living, there is no evidence whatever in support of evolution; but on the contrary, every fact and phenomenon cognizable by the senses strongly contradicts that theory. This will become more and more apparent as we proceed.

Chapter II

Breaks in the Continuity

As we trace in imagination the supposed course of evolution from its assumed beginning in undifferentiated matter onward and upward to the infinite diversities of the organic kingdom, we not only encounter difficulties at every step and in connection with every detail, but we also find certain gaps, deep and wide, for which evolutionists themselves can offer no definite explanation. The first and greatest of these is the gap between the living and the not living. The entire world of living creatures is assumed to have emerged, some time and somehow, and through "resident forces," out of the inorganic realm. Yet no trace of this marvelous process remains, and the inorganic world exhibits no progressiveness at all, no power or disposition to advance one hair's breadth.

The next gap is between the vegetable and animal kingdoms. If the latter, in its entirety, arose out of the former through gradual and infinitesimal changes, no trace of that marvelous development remains; nor can there be found in the vegetable kingdom anything from which the characteristic features of animal life could be evolved.

Next, we encounter the great gap between the vertebrates and the invertebrates; then that between the mammals and other vertebrates; then the gaps between each of the two million or so of distinct species of organisms and every other; and finally the immense gap between Man and the highest of the brutes.

In considering these great gaps, and the many lesser ones, it should be borne in mind that Evolution is set forth expressly as a theory of origins, that is to say, as an explanation of how all the infinite varieties of things, living, and non-living, came into existence. But origins, including those of the very broadest kind, are just what the theory conspicuously fails to explain. Thus, to begin with, the evolutionist makes no pretense that his theory can explain the origin of either Matter or Force. The existence of these he must take for granted, and attribute them to an Unknowable First Cause.

The Origin of Life

Going on further we come to creatures having that mysterious thing called Life. Does Evolution account for the origin of that? Quite the contrary; Darwin himself declared that spontaneous generation is "absolutely inconceivable." His co-discoverer, Alfred Russel Wallace, says that "the very first vegetable cell must have possessed altogether new powers"; and he adds, "Here we have indications of a new power at work." Huxley admits that there is no evidence of any link between the living and the not living, and other leading evolutionists could be quoted to the same effect. So, just where an explanation of the origin of a new and extraordinary thing is needed, Evolution--that great theory of origins--completely breaks down. Matter and force do not account for the origin of life. Therefore, Darwin had to accept the truth of divine fiat to explain it. He seems, in accepting this truth, to seek, using fine language, to disguise the fact that it is fatal to his theory. Note his words: "There is a grandeur in this view of Life, with its several powers, having been originally breathed by the Creator into the first forms or into one." If so, then is there not the same "grandeur" in the view of the direct action of the Creator in the origin of every substance and species? Mr. Darwin admits (because he must) that there is nothing "unscientific" in assuming the direct intervention of the Creator in originating the first living forms; and if so there is nothing unscientific in assuming His intervention to create all living species.

The Origin of the Animal Kingdom

Going still further, we come to the animal kingdom, whose species have powers (as locomotion, feelings, etc.) not possessed by the vegetable. What has Evolution to say as to the origin of that? Not a word. Here again, Mr. Wallace admits that the advance from the vegetable to the animal kingdom is "completely beyond all possibility of explanation by Matter, its laws and forces. It is the introduction of sensation or consciousness, constituting the fundamental distinction between the animal and vegetable kingdoms."

Thus, with respect to the origins of the major divisions of nature, the theory of Evolution is a confessed failure. It cannot even pretend to account for them. This fact will be emphasized when we come to point out, later, that the foremost evolutionists, including Spencer, Huxley, and Romanes, before their death, utterly repudiated the Darwinian theory of the Origin of Species. It would be difficult or impossible to find a naturalist of the first rank who would support that theory today. Haeckel alone, of the older naturalists, stood for its defense; and he was utterly discredited because of his audacious and unscrupulous conduct in forging pieces of evidence to support the theory. But we would at this point ask, what is the value of, and what credence should be given to, a Theory of Origins which admittedly is unable to account for the origin of Matter, Force, Life, the Animal Kingdom, and the Species?

Other Origins

Passing on, in our general survey of nature, and without further reference at present to the Origin of Species (of which we have more to say later), we come to the Vertebrates, that is to say, that large and superior order of living creatures which have a back-bone. Does Evolution give us any explanation of that? None whatever. While the difference between the vertebrates and invertebrates is not so conspicuous and notable as that between the living and not-living, or that between the vegetable and animal kingdom, yet Evolution is just as impotent to explain the one as the other. There is not the slightest evidence to show that creatures having no backbone "evolved" the many species of vertebrates, nor is there even a plausible suggestion as to how such a thing could be brought to pass.

Looking further, we come to the large and important class of Mammals, that is to say, species whereof the female imparts nourishment to its young from the breast. Certain species of vertebrates--the birds, reptiles and fishes--have not this peculiarity, nor anything approaching it. These, however, are far inferior to those creatures which have the nourishing breast. So, we ask again the question: Does Evolution account for it? And again, the answer must be in the negative. There is no connecting link between the two classes; nor are there any groups of non-mammals that are reaching out to enter the great class of Mammalia.

We would at this point dispose of an unwarranted inference which is often urged (in the total absence of proofs) in support of the theory of Evolution. That inference is that because there are many species which have features in common--as back-bones, and nourishing breasts--those species must have had a common origin. Resemblances are supposed to point to an ancestor common to all. But the inference is without a warrant. Such resemblances are just as consistent with the dogma of Creation as with the theory of Evolution. Resemblances are to be expected in the works of an all-wise Creator. For when He has devised a contrivance, as a backbone, to serve a certain purpose, He would inevitably use the same device in whatever place that purpose was to be served, with only such modifications and adaptations as the varying needs of individual species might require.

We need not continue any further, for our present purpose, our general survey of the field of the organic kingdoms of nature. We deem it enough under this heading to say that, in not one of these orders and species, and in not one of the countless billions of organisms comprised in them, has there ever been seen the slightest tendency to advance, or to depart from its type. On the contrary, there is found in every living creature the most stubborn and unconquerable determination not to evolve. The whole universe, therefore, and every member of it, is a witness against Evolution.

Furthermore, if Evolution were the law of progress of the universe, it is manifest that there would be no species or other lines of division. There would be only individual forms, shading imperceptibly one into another, each in the process of becoming something else, so that classification would be an impossibility. The world that lies before us, composed of clearly marked divisions, orders, classes, species, all sharply defined and separated one from another by impassable barriers, is just the opposite of such a world as the supposed law of Evolution would produce. We can, therefore, summon heaven and earth, land and sea, and all the hosts of them to bear witness, that Evolution is a myth.

Permanence of the Species

The matter of the permanence of species deserves special consideration. It is admitted on all hands that there is no Evolution in the individual organism--but that the contrary rule holds everywhere. For the individual comes into being suddenly, matures quickly, reproduces, and suddenly ceases to be. The evolutionist, however, claims that it is the species, not the individual organism, that has come into existence through Evolution. The species, says he, is governed by the "law" of Evolution, though (strange to say) the individuals which compose the species are exempt from it.

There is, however, a serious and obvious flaw in the logic which would distinguish thus between the individual, and the species to which it belongs; for the species cannot exist apart from the individuals composing it, any more than a river can exist apart from its water. The species is merely an abstraction; and there is obviously no way a species can evolve, except by the evolving of all the individuals composing it. Strictly speaking, and for the purposes of a discussion like this, "species" do not exist. What exists in nature are simply innumerable individuals each having their own life. Individuals which have a life of the same sort are said to constitute a "species." Therefore, evolutionary changes, if they take place at all, must needs begin and continue in individuals.

It follows that, if there be no inherent tendency in individual organisms to depart from their ancestral types, there could not be any development of new species. If, on the other hand, the immense number of existing species did come into their present state of being through evolutionary changes, effected by resident forces (as distinguished from the act of a Creator from without) then we should find no distinct species of plants and animals, but a very different state of things; for, instead of definite and persistent types, we should have a confusion of nondescript individuals, each in the process of becoming something different from what its ancestors were.

Evolution assumes that all things in the organic world are endowed with two opposite and mutually antagonistic tendencies, first a tendency to depart from its ancestral type under the influence of changes in "environment," and second, a tendency to hold tightly all its peculiarities and to transmit them to its offspring. These two tendencies could not exist in the same creatures. The former is purely imaginary. It is contrary to all the observed facts of nature. For, so far from there being any tendency on the part of individuals to depart from the ancestral type, and so far from there being any evidence of "resident forces" in them, impelling them to do so, the fact is--always and everywhere--that individual organisms evince a most stubborn tendency to cling to the ancestral type, despite all influences to the contrary.

This important fact can be stated very strongly; for scientific men, like Luther Burbank, have sought by every conceivable means to develop new species. But, notwithstanding some remarkable results in the way of "varieties," it has been found (1) that the barrier of species cannot be crossed, (2) that every "variety" produced artificially, if left to itself for a few generations, reverts to the original type. In a word, what we find in each and all the thousands of species of living creatures is, perfect obedience to the primal law of their being, given to them by their Creator when He said, "Let the earth bring forth grass and herb yielding seed, and the fruit-tree yielding fruit after his kind * * * and the living creature, after his kind, cattle, and creeping thing, and beast of the earth, after his kind; and it was so" (Gen.1:11; Gen.1:24). It was "so" then, and beyond all question it is "so" now.

Here we see that Evolution comes into direct collision both with the facts of nature and with the statements of the Word of God.

Development of "Varieties"

A "species" may embrace many distinct "varieties," and man has indeed been able to produce artificially many varieties of existing species. But it is always necessary to maintain by artificial means the modifications thus produced, else the individuals speedily revert to the original ancestral condition. Thus, upon consideration of these modifications of type, it is found that, so far from lending any support to the theory of Evolution, they furnish a strong argument against it. For it is essential to that theory that modifications when of advantage to the possessor, should become fixed in the family, and be carried forward to all succeeding generations. But what we find in actual experience is just the reverse.

Moreover, while varieties without number can be easily produced, it has been found impossible, even in a single instance, to cross the line of species. Thus, we see many varieties of dog. The canine species include the great shaggy St. Bernard and the diminutive smooth-skinned terrier. But in every case, it is a dog and is recognized by his fellow dogs as such. No amount of breeding, or crossbreeding, could ever make him anything but a dog.

Indeed it is demonstrable that the species are absolutely fixed; and that so far from there being a general tendency on the part of all animate creatures to depart from the ancestral type, there is, on the contrary, found to be an invariable and inexorable law, which absolutely forbids such departure. Since we regard this fact as fatal to the Darwinian theory of the origin of species, we will give the explanation of it in the words of a famous evolutionist, Mr. Huxley, who says:

"If you breed from the male and female of the same race, you, of course, have offspring of the like-kind; and if you make the offspring breed together, you obtain the same result; and if you breed from these again, you will still have the same kind of offspring. There is no check. But if you take members of two distinct species, however similar they may be to each other, and make them breed together, you will find a check. If you cross two such species, then, although you may get offspring in the case of the first cross, yet if you attempt to breed from the products of that crossing (which are what are called hybrids) that is, if you mate a male and a female hybrid, then the result is that in ninety-nine cases out of a hundred you will get no offspring at all."

We need not inquire the explanation of this, though Mr. Huxley says, "the reason is quite obvious in some cases"; for the fact is admitted on all hands.

Now what, we would ask, is the inference to be drawn from this fact? Certainly, it follows that the evolution of one species from another is an impossibility; so that, at this point again, the theory breaks down completely. Indeed, we can read as much between the lines of the admission which Mr. Huxley himself is constrained, though with manifest reluctance, to make. He says:

"After much consideration, and assuredly with no bias against Mr. Darwin's views, it is our clear conviction that, as the evidence stands, it is not absolutely proven that a group of animals, having all the characters exhibited by 'species' in nature, has ever been originated by selection, whether artificial or natural." And again; "Our acceptance of the Darwinian hypothesis must be provisional so long as one link in the chain of evidence is wanting; and so long as all the animals and plants certainly produced by selective breeding from a common stock are fertile with one another, that link will be wanting."

Later on, Mr. Huxley definitely rejected the Darwinian theory, as we will point out hereafter.

Reproduction

We have referred in the foregoing pages to the power, inherent in all living creatures, to reproduce their kind. This universal fact, which obviously is essential to the continuance of every species, raises the important question, how did the power of reproduction originate? It is evident that the very first (as well as all subsequent) organisms must have possessed this marvelous power. Whence then did it come? Manifestly it could not have arisen by a gradual process of Evolution; for the very first organisms must have had it in the same perfection as their offspring. Here again, the doctrine of Creation appears to great advantage in comparison with the defective theory of Evolution; for, as a prominent part of the inspired description of Creation, are the words: "Grass, and herb yielding seed after his kind" etc. Those words fully account for the power of reproduction possessed by all living creatures.

In concluding under this heading, we want to say that it would suffice to put the case for Evolution entirely out of court that there should be found no evidence sufficient in character and amount to establish it. But the case against it is far stronger than that. For even those who give no weight to the testimony of the Bible on this point, have to admit that there are no observable tendencies on the part of any one of the billions of living creatures to depart from the ancestral type, but that, per contra, where variations have been produced artificially, they are but slight in character, and the tendency is invariably to go backward and not forward. This is a strong disproof of Evolution.

Chapter III

Science not an Authority

In this chapter, we propose to examine the reasons usually advanced in support of the theory of Evolution. Those reasons relate entirely to Organic Evolution or the Origin of Species and the "Descent of Man"; for there is (so far as we are aware), no pretense that any facts are known from which the theory of Cosmic Evolution could be inferred.

It should be pointed out, before entering upon this examination, that it is easy to impose upon most people by an appeal to "Science" as an authority. Thus, we often hear it said, "Science has discovered this," or "Science tells us that," as if the matter were thereby conclusively settled. But it would be well to ask, who is "Science"? and where does he live? And how comes he to know these things? The fact is there is no "Science" in this sense. It is true that a few capable men have attempted to explore the field of Nature in various directions, and have ascertained a fact or two, to which they have added a thousand guesses. But they have left a million questions unanswered, without which no safe conclusions can be drawn. It is the commonest thing for "Science" to contradict one day what it most positively asserted the day previous; so that, in view of the existing state of complete scientific ignorance on the subject of origins, it would be absurd to accept as true any statement on that subject in the name of "Science."

A few years ago, Mr. Thomas A. Edison, commenting upon the boasted progress of Science, said that if the same rate of progress were maintained for the next two thousand years, mankind might then be in a position to begin to draw conclusions.

Reasons Given in Support of Evolution

In examining the reasons that are commonly given in support of the theory, we shall select those which are deemed, by its advocates, to be the strongest. These are (1) the changes which are observed to occur in the embryo of the human species from its first beginning to its full development, which changes are assumed to be, in their order and character, a recapitulation of the changes through which the species itself is supposed to have passed in the course of its development; (2) the succession of living forms in time; it having been ascertained from geological researches, that the more simple forms of life are, generally speaking, found in the lower strata of rocks, and those more complex higher up.

Embryology

We put this argument first because (a) it is generally deemed the strongest, and (b) it is from out of the studies of changes in the embryo (embryology) that the idea of Evolution sprang. So we have now the opportunity to examine the theory at its point of origin.

The argument from embryology consists of two suppositions, for neither of which is there any proof whatever. First it is assumed that the human species did evolve by gradual changes, passing from a simple unicellular creature, such as the Amoeba, through successively higher species until it became Man; and second, it is assumed that the human embryo passes through the same changes in its prenatal history of about nine months. Manifestly we have here no proof of Evolution; for in this argument, Evolution is taken for granted. It cannot be possibly known whether the changes of the embryo are a resume of the history of the development of the species until it is known what that history was. Therefore, we are thrown back upon the question, can Evolution be certainly inferred from the changes of the embryo?

First, let it be observed that there is no proved or necessary relationship between the growth of the embryo and the history of its species. If indeed the embryo does perform, in the short space of nine months, the stupendous feat of changing from Amoeba to Man, passing swiftly through all the intervening species, it would be a most miraculous and supernatural thing, whereof it was vain to seek an explanation in the sphere of nature. Evolution, however, is exceedingly slow. It demands millions of years to effect slight changes. It denies and excludes the miraculous from the sphere of nature. It cannot, therefore, assume a prodigious miracle in its own support. The supposed transformations of the embryo tend not in the slightest to prove the truth of the theory. Those changes, like all others in the history of a living creature, belong in the category of the mysteries of life, concerning which science has been able thus far to give no explanation whatever. "The way of the tree of life" (Gen.3:24) has been effectually kept from all prying investigators.

But let us go deeper into the subject, and ask, what are the changes which take place in the human embryo? And do they really constitute an evolution? Upon pressing this inquiry we find first, that the likeness of the human embryo to that of other creatures at different stages of its growth is a mere superficial resemblance; for even the evolutionist would not pretend that there is any essential likeness between them; and second, that even in those superficial and transient resemblances the growth of the embryo does not go through the stages of the supposed development of man. These objections are fatal.

1. In a case of this sort, superficial resemblances count for nothing; for beneath them, there are, in fact, vital differences. The human embryo is, at all stages of its growth, the human embryo. It is at every stage, essentially different from that of the worm, the fish, and from that of every other mammal. Prof. Fairhurst says:

"It is evident that while all eggs, from that of the sponge to that of man, may seem to be alike in structure, they are really as far apart in their essential nature as are the fully developed sponge and the full-grown man. Taking the embryos of man and fish the argument of the evolutionist is as follows: The embryos of man and fish, at a certain stage of development, are closely alike in appearance; therefore, man and fish had a common ancestral origin. The conclusion which the evolutionist draws is based upon a mere seeming and very transient resemblance, while the fact that the two embryos are essentially unlike is shown by the vast distance apart at which they arrive by development. It is true that the embryos of vertebrates look much more alike than do the adults, and that the eggs are still nearer alike in appearance than are the embryos; but I insist again that the embryos are no nearer together in essential structure than the adults. The egg which can be developed into a man is just as different in nature from the egg of a fish, as the man is from the fish. The eggs are essentially unlike. The essential qualities of eggs are beyond the power of the microscope to reveal. The human embryo is produced by human beings only; and whatever may be its microscopic appearance, it is at every stage of its development strictly human. Embryology, as applied to Evolution, fails in that it deals only with the surface of things."

Thus, the strongest argument of the evolutionist breaks down completely for the reason that the facts are the reverse of what his theory calls for.

2. Furthermore, even the superficial changes of the human embryo do not represent anything like the complete line of the supposed human ancestry. Prof. Fairhurst says that the entire first half of the history of Evolution is not even hinted at in the epitome (Organic Evolution Considered, p. 147). Further he says:

"There are radical differences between the embryos of vertebrates and invertebrates. Worms and other articulates in embryo lie doubled backwards around the yolk, while all vertebrates are doubled in the opposite direction. According to the theory that the embryonic condition is a recapitulation of the stages of organic evolution, this fundamental fact of invertebrate embryology ought to have been preserved by the vertebrate. Evolution gives no account of this reversal of position by the vertebrates."

There are other gaps in the succession of changes through which the embryo passes; but it is needless to speak of them. Enough has been said to show that the argument from embryology is not only a far-fetched inference, but that the facts are the reverse of what the inference calls for.

It is of interest to note that Dr. Romanes, one of the most extreme of evolutionists, declared the facts of embryology to be "the most important of the lines of evidence" in support of Evolution. While holding those views he wrote strongly against the Bible doctrine of Creation, and against supernaturalism in general. "But later he changed his views entirely, and died in 1894, confessing his faith, not only in the providence of God, but in the deity of Christ." (Fairhurst: Theistic Evolution, p. 11.)

Succession of Species

The evolutionist also appeals, in support of his theory, to the fact that the fossils preserved in the sedimentary rocks indicate that the various species did not come into existence all at once, but successively; and that (generally speaking) the simpler forms came first into existence, and the more complex later in point of time.

To this argument the obvious answer is that the fact of the successive appearance of the several species does not tend in the least to prove that the later was derived from the earlier by process of evolution, or by any other process. The succession of the species can be explained by Creation, as well as by Evolution. In fact, the record of Creation in Genesis 1, declares that vegetation first appeared on earth, then fishes, then birds, then land animals, and finally Man. The geological remains show the same order. Manifestly then the argument from succession of species lends no support whatever to the theory we are discussing.

But we can go further than this; for when the details of the geological records, as presented by the science of paleontology, are examined, it is found that they bear heavily against the theory. This is conceded by the very foremost evolutionists, insomuch that, to escape the force of the paleontological proofs, they are driven to the pitiful expedient of supposing that there have been millions of extinct species and transitional forms which have left no trace of their existence, and that if by any means the lost evidence could be recovered, it would prove their case.

The fact is that an enormous mass of evidence has been accumulated by means of geological researches. Here we have the footprints of the distant past, the records of the periods which would certainly be rich in the pieces of evidence of the evolutionary origin of the various species, if such were indeed the nature of their beginning. The evolutionist examines this great mass of facts and finds nothing which supports his theory, but much to the contrary. His only comment on the situation is that Nature has, with invidious discrimination, destroyed the great bulk of the evidence, including every trace of the operation of Evolution, and every one of the thousand billion variant forms which must have existed, and has preserved only such pieces of evidence, and those in great abundance, as oppose his theory. It may be said of this explanation that it is even harder to understand and to accept than that which it purports to explain--the absence of all trace of a "law" which is said to have operated universally and from the very beginning of time. The great god, Evolution, is indeed as difficult to locate or find a trace of as the Olympian Zeus.

Imagine a litigant in court upon whom rests the burden of proof. He insists that the averments of his declaration are true and demands a verdict in his favor; but he has no proofs to sustain his allegations. In fact, all the evidence presented in court is against him. He demands, nevertheless, that judgment be rendered in his favor upon the supposition (a) that volumes of proofs, which once existed, have been destroyed, leaving no trace; and (b) that if those proofs could now be produced they would be found to be in his favor! Such is the absurd plight in which the theory of Evolution now finds itself, as matters stand at present.

As to this important feature of the discussion it is enough to say that, considering the great mass of fossilized remains which have been collected from every stratum, and from every part of the world, the presumption is that, if the records were complete, the parts now missing would confirm what we have.

Species Appeared Suddenly

The first fossil remains of organisms are found in the Primordial period. Le Conte says that in it are found "the representatives of all the great types of animals, except the vertebrates." Thus, according to the evidence (which, by Le Conte's statement, is massive in quantity and clear in character), numerous highly organized creatures--about half the entire animal kingdom in fact--came suddenly and virtually simultaneously into existence. Of their supposed progenitors, of whom, if Evolution is true, countless billions must have existed, not a trace survives in the earlier formations. This is more impressive because those earlier formations are estimated to have occupied about half the entire period of geological time.

(In this discussion we are giving the evolutionist the advantage of supposing, for the purpose of the argument, that his theory of immense ages of geological time is correct. That theory is, in fact, quite destitute of supporting proof, and we wish it to be understood that we do not accept it as true.)

Great Gaps Between Species Existed from the First

Another very striking fact which this earliest record of living creatures presents is that, "from the very beginning the great gulfs which separate the animal kingdom into sub-kingdoms and classes existed then and have continued till the present time" (Fairhurst). Considering that the interval from the Primordial period until now is estimated by physicists at fifty millions of years, we have in this fact of stability of the species a conclusive proof that Evolution is a myth.

Another striking fact, to which these records bear witness, and which is fatal to the theory, is that every species, as it suddenly appears, has its complete organism; that is to say, it is fully developed in every feature of its structure, however complex.

If, therefore, we place ourselves in imagination in the Primordial period, amidst the immense number of varieties of living creatures then existent, whether we look back into the past, or forward into the future, we see that Evolution had nothing whatever to do with their origin or development. In one direction we see no long ancestral line from which they were gradually evolved; for the species, like each individual member thereof, came into being suddenly. This may be termed "negative" evidence. But such evidence is sometimes conclusive, as when a thorough exploration of an island reveals no remains whatever of man or human implements, it may be concluded with certainty that it was never inhabited by man.

But on looking forward the evidence is positive, as well as conclusive. For the very same species found in the Primordial era, and appearing suddenly, are in existence today without substantial change of structure or habit of life. Evolution requires, and of course would produce, life-forms quite flexible and plastic, structures such that every part of every organ and surface would be liable at all times to variation and would be subject to change whenever a change of environment occurs. But we find, on the contrary, absolute rigidity of both structure and habit. On this evidence, we are bound to conclude that living creatures originated in a manner very different from that assumed by the evolutionist.

As has been already stated, the earliest geological remains of organisms show lower and higher forms of life existing side by side. Now, according to Evolution, the former would be the progenitors of the latter; and upon that supposition, there must have been already at that early period an immense evolutionary advance, which would imply that such lower forms were exceedingly progressive in character. But this supposition (and with it the entire theory) is completely negated by the fact that those self-same forms have persisted without change to this very day. Instead of being progressive, as Evolution demands, they are proven to be absolutely unprogressive. Every one of those million forms is a venerable witness (50 million years old, if our geologists are right) against the theory of Evolution.

What reply has the evolutionist to these facts? Worse than none. Mr. Huxley, one of the ablest of them all, and one who openly devoted his great talents to the destruction of faith in Divine revelation has faced these facts in his address to the Royal Geological Society in 1870. He puts the question thus:

"What then does an impartial survey of the positively ascertained truths of paleontology testify in relation to the common doctrines of progressive modification (i.e. Evolution), which suppose that modification to have taken place from more to less embryonic forms, from more to less generalized types, within the limits of the period represented by the fossiliferous rocks?" And he answers the question by saying, "I reply, it negatives those doctrines; for it either shows us no evidence of such modifications, or it demonstrates such modification as has occurred to have been very slight. The significance of persistent types, and the small amount of change which has taken place even in those forms which can be shown to have been modified, becomes greater and greater in my eyes, the longer I occupy myself with the Biology of the past" (quoted by Th. Graebner, in "Evolution").

The Fragmentary Character of the Record

The disappointed evolutionist pleads the imperfection of the geological records. In order that his theory may not be dismissed for lack of proof, he asks us to believe that much of the pertinent evidence has and that what has been lost was in favor of his theory. But Le Conte says: "We think the fragmentariness of the geological record has been overstated." And the Duke of Argyll, in his Organic Evolution Cross-Examined, shows conclusively that, in certain periods, the plea of the imperfection of the record is completely ruled out.

"There are," says he, "some tracts of time regarding which our records are as complete as we could desire. In the Jurassic rocks, we have a continuous and undisturbed series of long and tranquil deposits, containing a complete record of all the new forms of life which were introduced during those ages of oceanic life. And those ages were as a fact long enough to see not only a thick (1300 feet) mass of deposit but also the first appearance of hundreds of new species. These are all as definite and distinct from each other as are existing species. No less than 1850 new species have been counted, all of them suddenly born, all of them lasting only for a time, and all of them in their turn superseded by still newer forms. There is no sign of mixture or of confusion, or of the infinitesimal, or of intermediate variations. These 'Medals of Creation' are all, each of them, struck by a new die, which never failed to impress itself on the plastic materials of this truly creative work."

Could it be shown that but only one species originated otherwise than by slight modification of the structure of pre-existent species, that would suffice to overthrow the theory of Evolution. But the science of paleontology presents us with clear pieces of evidence of thousands of species coming suddenly into existence; and on the other hand, there is not the faintest indication that there was ever a species that came into being in any other way.

We have, therefore, found that what evolutionists put forward as the strongest proofs of their theory--Embryology and Paleontology--yield, when closely examined, convincing, indeed conclusive, the evidence against it.

Chapter IV

Specific Objections to Evolution

Evolution undertakes to account for every part of every living organism, by progressive modifications caused by resident forces. Hence it is not an exaggeration to say that every organ and member of every living creature supplies us with an objection to Organic Evolution. Our difficulty, therefore, is not for lack of illustrations, but rather which to select from the number available. Several out of the many at our disposal will suffice to show how completely the theory breaks down when we leave the realm of vague generalities and bring it to the test of concrete facts.

Before applying this test we should perhaps state that, according to the Darwinian theory, every organ was evolved from what was originally a very slight variation (due to accident or other cause unknown), which variation, because it proved useful to its possessor, was transmitted to its offspring; and it is supposed that the departure continued through many generations until at last, it became an organ--such as an eye, an ear, a wing, with a distinct and valuable function. Thus "Natural Selection" attempts to account for the preservation of certain variations from the original stock, but not for their production. The main point of the doctrine is that only such accidental variations are preserved as are advantageous to their possessor. We cannot state the theory more definitely because its authors themselves are utterly unable to suggest how Natural Selection worked in any concrete case--as, for example, in evolving the wings of fowls and insects. Mr. Darwin says: "Our ignorance of the laws of variation is profound. Not in one case out of a hundred can we pretend to assign any reason why this or that part has varied." Thus we are left, as Prof. Fairhurst says, "in almost total darkness as to the cause of the most important factor in Organic Evolution."

The evolutionist leaves us to think out for ourselves how the limitless number of diversities of organs, members, instincts, etc., in all the millions of living species, came into being. We see in all of the specific organs upon which their existence or welfare depends. Natural Selection tells us that, at a time far back in the past, their ancestors had none of those organs, not even those that are vital. But it does not, nor does it attempt to, trace the development of a single organ, or tell us what the intermediate creatures were like, or how they lived during the long stretches of years during which those vital organs were being evolved. The question is: How did many generations of species live without organs whose functions are vital? That is a hard question even for an evolutionist.

Here then we are in a position to state an objection to which, so far as we are aware, no reply has ever been made. It is this: Inasmuch as the evolution of an organ, such as the wing of a fowl, would require many centuries of time, and many generations from parent to offspring to bring it to a useful stage of development, how is it possible to account for its preservation during the long period when it was an undeveloped and useless appendage? Natural Selection purports to account for the preservation only of such variations as are useful to the possessor in "the struggle for existence." The facts of nature force that limitation upon the theory, inasmuch as the existing organs and members are such as are of some use or advantage. The theory cannot admit of the perpetuation of useless organs and structural features, for such does not exist. But, upon looking closely at the matter, we perceive that every organ--such as an arm, an eye, an ear--however advantageous when fully developed, must have been preceded (if the theory be true) by an exceedingly long period during which it would have been not merely useless, but often a positive disadvantage. It follows that Natural Selection, by its own necessary limitations, cannot account for the development of any organ which must need pass through a period of non-usefulness. Hence the theory breaks down completely.

The Wings of Fowls

Consider, for example, the wing of a fowl (an illustration used most effectively by Prof. Luther T. Townsend and quoted in Dr. Leander S. Keyser's Contending for the Faith). Here is a very highly organized structure, certainly most important to its possessor. It is a wonder of design, and the very perfection of workmanship in every detail, down to the tiniest feature of the smallest feather. Whether we regard the design, or the construction, or the material, we see perfection in each and all. But we find on the back of every bird not one wing, but two, practically identical in every feature. Moreover, they are symmetrically placed, and in the most advantageous position for the purpose for which, upon the theory of Creation, we should say they were intended. But, according to Evolution, those wings must have been developed each quite independently of the other, from what was at first an accidental hump or protuberance on the back of a featherless reptile. They must, moreover, have been perpetuated, with steadily progressive development, keeping pace with one another, through the progeny of countless generations, during all of which time, these unnatural excrescences would be, not an advantage, but decidedly an encumbrance to their possessors. But this could not go on under the "law" of Natural Selection; for that "law" tolerates only the fostering of useful variations. Hence Natural Selection would quickly destroy such variations. But conversely, the wings of the fowl destroy Natural Selection. Evolution cannot account for wings, either by Natural Selection or by any other supposed method of working. Many able evolutionists have admitted this (Herbert Spencer among them); yet they cling to Evolution, notwithstanding the impossibility of proposing a method by which it could work. Is it because they cannot bear the alternative of recognizing Creation and the Creator?

The Bat and the Mole

We would cite in this connection the instructive case of the bat, quoting from Prof. Th. Graebner: "The bat," says he, "is another highly specialized animal.

"In many respects, it resembles the mole; but its hands are enormously expanded, and the exceedingly long fingers are connected by a soft membrane, making a most serviceable wing. Is it not extremely likely, assuming the development theory to be true, that the mole and the bat sprang from a common ancestor? And was not that ancestor probably a wingless mammal? How then did the bat acquire his wings? Did he attempt to spring into the air to seize a passing insect, reaching out his forepaws to catch it? And did those paws gradually become enlarged until, after some generations, they became real wings? If so, what happened in the meantime to those connecting links whose wings were but partly developed? A bat with wings only half-grown would be a helpless creature and would surely perish. There is no middle ground. If the ancestor of the bat was a terrestrial creature, with limbs fitted for walking, then it must have given birth to a full-fledged bat, fitted for flying. There could have been no middle stage; for such a creature would have been helpless and must have perished.

"All this applies with equal force to the diversified and often highly complex structure of plants. As the organs of the various plants are now constituted, they serve most admirably their respective purposes. Given a slight change, an un-development, and the individual would perish. But such undeveloped stages must necessarily have occurred in the history of every life-form on earth if a change through slow adaptations is to be accepted as a hypothesis to account for their present form. To our mind this matter of rudimentary structures presents an insuperable obstacle to acceptance of the evolutionary hypothesis, even on scientific grounds."

We have thus far considered only the subject of wings, seeking to imagine how those wonderful organs, so vital to their possessors, could have been evolved. But manifestly whatever organ or member, external or internal, of whatever creature we might select, it would be equally impossible to trace any line of development for it, whether by Natural Selection or any other method of Evolution that has been proposed. It is obvious that humps, excrescences and other abnormalities, are blemishes; and the more they might be developed, short of acquiring a new function (as sight, hearing, flying) they would be a great disadvantage to their possessors. Such abnormalities, moreover, do not tend to reappear in offspring. On the contrary, they tend to disappear. A whole race of men has practiced the rite of circumcision for nearly four thousand years, and at the same time have refrained from outside marriages; yet never was a child born already circumcised.

If, however, the perpetuation of such abnormalities was indeed the law of nature, then there would be no recognizable species. All individuals would be undergoing changes, both internal and external. In such case we should see humps, protuberances and the like, on various parts of different creatures, in various stages of progress towards whatever chance, or "resident forces," might ultimately determine--legs, arms, wings, horns, tails, trunks, tusks, or some other and novel sort of organs or members, of the nature whereof we could form no idea in advance of their complete development. That is what we would see if Evolution were true. If then we see nothing of the sort, it is because Evolution is a delusion.

It is appropriate also to ask, when, under the supposed process of Evolution, would a developing organ or member reach completion? How would the "resident forces" know when to stop its progress? Could it ever be said, in any case, that an organ was finished? Would not progressive changes be always taking place in every part of every organism? Certainly, if the world of living creatures were indeed left to the blind control of unintelligent "resident forces," it would be a world of more vagaries, monstrosities, and abnormalities than was ever pictured by a delirious brain, or by the disordered imagination of an opium-eater.

The Water Spider

Let us now consider the case of the water spider and ask ourselves if there be any conceivable way in which its peculiar organs, instincts, and manner of life, could have been derived, by Evolution, from others of the spider family.

Like other spiders the water spider is an air-breathing animal, yet, unlike other spiders, it lives underwater. How did it evolve the extraordinary changes in its organism, and in its habits of life, whereby it acquired first, its set purpose to live underwater; and second, its special organs and instincts whereby it is enabled to give effect to that strange purpose, and to live, thrive, and rear its young in such an unnatural environment?

Of course, if the water spider was always a water spider, and was, by its Creator, endowed with just the organs and instincts that are suited to the manner of life appointed to it, the matter is very simple and intelligible. But we are inquiring how the water spider and its ways could have come about through Evolution. Surely those who press that theory upon their fellow-mortals, and who ask them to cast aside the belief in Creation and the Creator--with all that that involves--should at least be required to tell us how Evolution worked or could have worked in such a case. Was ever such a thing heard of, as that we should be asked to believe, on the ground of "reason" and "Science," in a thing so preposterously unreasonable that the imagination can conceive of no possible way in which it could be accomplished?

Upon examining the water spider, and acquainting ourselves with its ways, we find that its body is covered with hairs in such a way that it does not become wet when in contact with water. In order to live underwater, and rear its young there, it must construct a water-proof cell, capable of containing enough air for breathing purposes; it must have means for renewing the supply of air from time to time, and it must have the instincts to guide it in the performance of these necessary operations. And we may confidently add that the very first water spider must have been fully equipped for the purposes indicated. It spins under the water an egg-shaped envelope, open underneath for entrance and egress. This envelope, which is waterproof, is securely attached to some object so that it will remain submerged. Having constructed its house, the little creature next proceeds to fill it with air. For this necessary operation its hind legs are covered with hair and are so constructed that they can take hold of a large bubble of air, and carry it down into the water, and to the opening of its house. There the air is released, and it rises to the top of the envelope, expelling the corresponding quantity of water. This operation is repeated until the cell is sufficiently filled with air. The eggs are then laid in the upper part of this house and are surrounded by a cocoon.

It is manifest that this extraordinary manner of life, and the highly specialized organs, which are vital to it, could not possibly be the outcome of a long and slow process of development. Before the life of a water spider could even begin, it must be equipped with, first, the means for secreting a water-proof material; second, means for spinning that material into a water-tight cell; third, protective hairs to keep it from becoming wet; fourth, the peculiar apparatus for filling its house with air; fifth, the several instincts which prompt the doing of these remarkable things.

That there is no trace of the evolution of the water spider (or of any other creature) is reason enough why the theory should be rejected. But we confidently submit that the facts briefly set forth above, and the conclusions which necessarily follow from them, constitute proof positive that Evolution is not only an impossibility but an absurdity.

Spiders in General

The subject of the origin of instincts will be further considered under our next heading. But while we have before us the subject of spiders, the following from Orton's Zoology will be of interest:

"Spiders are provided at the posterior end with two or three pairs of appendages called spinnerets, which are homologous with legs. The office of the spinnerets is to reel out the silk from the silk-glands, the tip being perforated by a myriad of little tubes, through which the silk escapes in excessively fine threads. An ordinary thread, just visible to the naked eye, is the union of a thousand or more of these delicate streams of silk. These primary threads are drawn out and united by the hind legs."

Here we find a marvelous coordination of special organs: (1) the silk-glands, capable of secreting a fluid which has the remarkable property of hardening upon exposure to the air; (2) spinnerets, having each more than a thousand perforations of microscopic size, without which the silk-glands would be worse than useless; (3) hind legs, having the wonderful function of forming the thousands of invisible filaments into a thread, without which function both glands and spinnerets would be a serious detriment to their possessor. It is simply impossible that these three organs should have developed gradually, and independently of each other, to the stage of perfection, in advance of which stage they could not co-operate in the slightest degree to the one end for which they all exist.

Let it be noted that if the spinnerets had but one aperture, or a dozen, or even a hundred, the liquid material would not have the required area of exposure to the air to the effect that instant solidification which is absolutely essential to the success of the entire operation. It required at least a thousand apertures to produce the desired result. Who knew, or could have known, the need for such a number of orifices? or could have formed them in a tube the size of a spider's leg? And in what imaginable way could several legs, intended for locomotion, be evolved into organs so radically different in function? It is not too much to say that those thousands of orifices are just so many witnesses that Evolution is a huge delusion, which has made foolish the wisdom of the wise and has exposed to deserved ridicule the gullibility of the brightest minds.

Organs and Instincts: The Bee-Hive

The difficulty of tracing a line of development along which any known organism could have come into being, or any of its special members or parts could have originated, is immensely increased when we take into consideration a highly specialized creature, such as the honey-bee, which is also endowed with unique instincts requiring for their exercise a corresponding unique structural organization. In such case, the theory must account, not only for the evolution of an exceedingly complicated mechanism, but also for the simultaneous development of equally complicated instincts, dependent upon that very mechanism, and impossible of being obeyed without it. And it has further to account for the preservation of both mechanism and instincts through the long era of inutility. And--to add one impossibility to another--we have here a case in which, not the life of the individual only but that of the entire community depends upon the exercise of those instincts and the functioning of that mechanism. Where and what were the honeybees during the centuries of time which Natural Selection would require for the evolution of those instincts and their necessary mechanism? Evolution attempts not to give an answer.

But the difficulties, in this case, have not yet been fully stated; for in the swarm of bees we find three distinct classes--queens, workers, and drones. Each of these classes is absolutely necessary to the life of the swarm, and each has structural peculiarities and instincts radically different from the other two. The workers, which are undeveloped females, constitute the largest and most important class. Their organic structure is highly specialized to fit them for the many and various operations they have to perform, and their instincts are correspondingly complex. How and from what could such a marvelously specialized creature have been evolved? The evolutionist can give no answer that is worthy of a moment's notice. But the wonder of this largest and most important class of the bee-community is that, both in organization and in instinct, they are diverse from both their parents; for they are the offspring of queens and drones. It is vital to the theory of Evolution that the characteristics of parents should pass to their offspring. But here is a highly organized creature which has an organic structure and a complex set of instincts possessed by neither of its parents! Whence then comes the honeybee? It does not transmit its wonderful characters to its offspring, for it has none. And if a worker-bee should lay an egg (as occasionally happens) the offspring is invariably a drone. Clearly then, the worker bees are not the product of Evolution; and their existence and renewal from generation to generation, from parents unlike themselves, is a standing contradiction to Evolution.

The Beaver

Prof. Fairhurst, in his able work already quoted, (Organic Evolution Considered) calls attention to the remarkable example of instinct manifested by the beaver. We quote:

"It lives in communities and constructs dams, sometimes as long as three hundred yards, stretching across shallow streams of water. These dams are built of sticks of wood, generally about three feet long and six or seven inches in diameter, which the animal cuts with its teeth. The sticks are put in the water and are held in position by means of mud, stones, and moss, which are placed upon them. The dams are ten or twelve feet thick at the base; and when the streams are wide the dams are made to curve upstream against the current, thus producing a structure better able to resist the force of the stream. The amount of labor necessary to construct a large dam is enormous. Moreover, it requires an incredible number of logs of wood and great skill in engineering.

"Near the dams, the beavers build their houses. Each house is about seven feet in diameter in the interior, and three feet high in the center. The walls are of great thickness. Each lodge is large enough to accommodate five or six beavers.

"The outside is plastered with mud and carefully smoothed, and the mud is renewed each year in order to keep the house in good repair. All the houses of the colony are surrounded by a ditch which contains water, and each lodge is connected by a passageway with the ditch.

"As a supply of food for the winter, the beavers store up many logs under the water, the bark of which they consume.

"Thus we find in this case an organized community, working for the common good, both in constructing the dam and the ditch, and also in storing up food; and then making special preparation for living in small groups by constructing their lodges and connecting them with the ditch.

"Here, we see highly developed instincts which look to the future good of the organism. The building of the dam, the digging of the ditch, the storing of the food, are all done to meet future emergencies. It is evident that the construction of the dam could not have been evolved gradually, for a dam must be of sufficient extent to be useful ere Natural Selection could act.

"Are we to presume that beavers experimented for countless generations, thereby building up the instinct which leads them to construct the dam? If so, upon what ground can we explain the preservation of the incipient instinct until sufficiently developed to be of practical use? In what way could they have known in advance that a dam would serve their good? Shall we assume that their instinct led them, in the first instance, to construct a dam, but not having had any experience whereby an instinct of that kind could be evolved? If the instinct existed without having been developed by experience, then we cannot account for it by Evolution"--And we may interrupt our quotation to say that the instinct must have existed in advance of the building of the first dam, else obviously it would never have been built.

"If evolved, then we must assume that the first dam made was of sufficient use to give its makers an advantage in the struggle for existence and that the instinct which led to its construction was transmitted to their offspring.

"Manifestly then, in accounting for the evolution of this instinct, we of necessity begin with an instinct that is already useful; and thus, we assume the existence of that for which we are trying to account. We are obliged to assume that, in a single generation, a beaver or colony of beavers was produced, which had a new instinct, sufficiently developed to enable them to build a useful dam; and that, in consequence of this, they themselves were the better preserved; and that the instinct was transmitted to the offspring. If all this could have happened in a single generation, it is evident that no question need be raised as to the possibility of future evolution.

"Besides this, the construction of a ditch for water around the several lodges required a different instinct, serving another purpose. Its evolution involves similar difficulties."

The examples considered above are not exceptional; for we could never exhaust the strange instincts of insects alone, of the origin of which it is impossible to account upon the theory of Evolution.

The question of the development of instincts, along with that of special organs, required for those peculiar instincts, and in their turn utterly useless without the latter, is a question which the evolutionist is unable to face. Mr. Darwin himself says there exist "cases of instincts almost identically the same in animals so remote in the scale of nature, that we cannot account for their similarity by inheritance from a common progenitor, and consequently must believe that they were independently acquired through Natural Selection" (Origin of Species, p. 226).

But Mr. Darwin himself realized that, to believe a thing so utterly unreasonable, and so contrary to all known facts and experience, would require credulity of a most uncommon sort; for he said: "Many instincts are so wonderful that their development will probably appear to the reader a difficulty sufficient to overcome my whole theory."

True enough. For in this we can thoroughly agree with Mr. Darwin. But inasmuch as Mr. Darwin was evidently himself aware of the incredibility of his theory, we wonder how he could expect others to accept it. What the whole extraordinary situation demonstrates most conclusively is, that there is no mind so capable of believing the incredible, as that which is pleased to call itself "the scientific mind," and that there is no person in the world so irrational as the "rationalist."

Chapter V

The Origin of Man

Of all the questions of origin that of Man is supremely important, and if, as we doubt not is the case, the doctrine of Evolution was inspired by the great "Spirit of Error," as a special effort in these last days to "blind the minds of them that believe not," then we may well conclude that his main object would be to discredit statements of Scripture which relate to the creation of Man. The words, "Let Us make man, in Our image, after Our likeness" (Gen.1:26), reveal a truth of fundamental importance. Against this foundation truth of Scripture (which is closely linked to that of Redemption by Him Who came in the likeness of Man) Evolution raises the monstrous and impious fiction that Man was made in the image and likeness of the ape, by means of an unbroken continuity of changes imperceptibly small.

If the explanation of the origin of Instincts presented difficulties which Evolution finds insurmountable, what shall we say of those powers and endowments of mind and spirit which distinguish human beings, and which mark the existence of a mighty chasm, deep and wide, between the highest of the brutes and the lowest of the human race! For it is not in his physical being, his body, that the special characteristics of man are to be found. Physically he is far inferior in strength and activity to many brutes. His bodily resemblance to the largest of the apes is seen at a glance; but that resemblance is superficial, and is easily accounted for, consistently with the truth of Creation. For, since Man has a physical being, and requires organs for locomotion, sight, hearing, manipulation, etc., in common with other animals, his physical makeup would, of course, resemble theirs in respect to those organs, with only such modifications as would be required by the differences in his physical manner of life.

The Characteristics of Man

The differences--immeasurably great--between the brute and the man lay beneath the surface, and have their existence in the regions of the soul and spirit, regions which, though so manifestly real, are yet so mysterious that even Man himself has no means to explore them, nor words to describe the simplest of their mysteries. In Man we find a creature who is self-conscious, who can reflect, reason, contemplate; who has the power of abstraction; who can comprehend general ideas; who can arrange his thoughts; who can communicate them to others by oral and written language; who has a sense of beauty; ability to enjoy harmonies of sound or color; a perception of right and wrong; a conscience; and above all, who has a capacity to know God. How vast are these differences! Who can declare their breadth and depth? Who would, unless infatuated by some mysterious delusion, or possessed by the spirit of mischievous error, compare with the chattering ape a being of whom even Charles Darwin says that he can "follow out a train of metaphysical reasoning, or solve a mathematical problem, or reflect on God, or admire a grand natural scene"?

"Missing Links"

The very lowest type of human being has all these marvelous capabilities in common with the highest. Between the two extremes, there are infinite gradations, merging one into the other, in such a way that not a line the thickness of hair could be drawn at any part of the scale. But, when we reach the lowest limit in that scale, and look from that point to the very highest of the brute creation, it is not across a mere "break in the chain of continuity" that we are looking, but across an immeasurable chasm. There is much talk about "the missing link." But such talk is nonsensical. Moreover, it obscures the facts of the case; for it is not a mere "link" that is missing, but ten thousand times ten thousand links. There is nothing in all the animal creation which answers to the mental, moral and spiritual nature of Man; or to his power of verbal expression of thought; nothing from which those marvelous and godlike attributes and powers could conceivably be derived by Evolution.

The low moral standing of savages does not lend support to Mr. Darwin's theory, though he appeals to it; for the lowest of them possesses a capacity for morals, as well as the most highly cultivated of men. According to Natural Selection, says Prof. Fairhurst, "Savages ought not to have any capacities except those that have been constantly in use, and that have been preserved because they have proved useful." But savages have, in common with all men, natural powers which enable them to appreciate moral distinctions, and to receive instruction in regard thereto, and to make progress in education in all directions.

It is related of Mr. Darwin that "When he sailed past certain islands in the Pacific, he found them inhabited by cannibals; but twenty-five years thereafter be found those very islanders converted to Christianity and enjoying the blessings of civilization. How many millions of years would it take," can any evolutionist tell us, "to convert a tribe of gorillas into God-fearing, man-loving, self-conscious beings, capable of believing that they possessed immortal souls," and that they were the objects of God's redeeming love and saving grace? (Fairhurst: Organic Evolution Considered.)

What these facts of common knowledge prove is that all men are akin to each other; that all are in common endowed with attributes God-like in character, bestowed from above, not derived from beneath; that they are wholly distinct from the brute creation; that by sin the moral nature of Man has been ruined; but that Man, in his most degraded condition, is capable of being regenerated and renewed under the potent influence of the Gospel of Christ. Thus, the pertinent facts of common knowledge, which are universal and the same in all the centuries of our era, are in perfect agreement with the statements of Scripture concerning the Creation, the fall, and the recovery of Man through Divine intervention, and are utterly opposed to Evolution.

Ancient Human Remains

We would deem it a waste of time to discuss in detail the human bones, found in various localities, and which have been put forward by the evolutionist--hard-pressed for proof--as being the remains of a type of Man somewhat nearer to the ape physically, than any now living. It suffices to say that the actual character of such fragmentary remains is, in all cases, more or less a matter of speculation; and that not a single human skull or other bone has ever been discovered that differs in any marked degree from corresponding parts of human beings now living. Furthermore, we have shown that the real problem of the evolutionist is, not to account for Man's physical being by Evolution (impossible as that is) but to account for the origin of his mental, moral and spiritual attributes which are his real distinguishing characteristics. As to those, the facts all bear witness that Evolution is a monstrous delusion--doubtless a phase of that "strong delusion," to which, according to the prophetic Scriptures, they of the last days were to be given over, who "received not the love of the truth that they might be saved" (2Th.2:10-12). It will suffice, on this branch of our subject, to quote a few passages from recognized authorities.

The great Virchow, one of the very greatest of chemists, an investigator, and anatomist of the first rank, says:

"We must really acknowledge that there is a complete absence of any fossil type of a lower stage in the development of Man. Nay, if we gather together all the fossil men hitherto found, and put them parallel with those of the present time, we can decidedly pronounce that there are among living men a much greater proportion of individuals who show a relatively inferior type than there are among the fossils known up to this time."

Honest evolutionists will not dispute this. Thus, the "Engis skull," found in Belgium, and gleefully hailed as that of the much sought "missing link," was conceded by Prof. Huxley to be "a fair average skull, which might have belonged to a philosopher, or might have contained the thoughtless brain of a savage." This Engis skull is supposed to be the oldest known up to now.

Again quoting Prof. Virchow:

"We seek in vain for the missing link. There exists a definite barrier separating man from the animal, a barrier which has not yet been effaced--heredity, which transmits to children the faculties of the parents.

"It was generally believed a few years ago that there existed a few human races which still remained in the (supposed) primitive inferior condition of their organization. But all these races have been the objects of minute investigation, and we know that they have an organization like ours, often indeed superior to that of the supposed higher races. Thus, the Eskimo head and the head of the Terra del Fuegians belong to the perfected types. All the researches undertaken with the aim of finding continuity in progressive development have been without result. There exists no man-monkey, and the 'connecting link' remains a phantom."

The above quotations are as given in Th. Graebner's work "Evolution: An Investigation and a Criticism," from which we also quote the following:

"No one has stated ascertained facts touching the origin of Man more succinctly and more clearly than Prof. Dr. Friedrich Pfaff, professor of natural science at the University of Erlangen. He shows conclusively that the age of man is comparatively brief, extending only to a few thousand years; that man appeared suddenly; that the most ancient man known to us is not essentially different from the now living man, and that transitions from ape to man, or from man to ape are nowhere to be found. The conclusion he reaches is that the Scriptural account of Man, which is one and self-consistent throughout, is true; that God made man in His own image, fitted for fellowship with Himself, a state from which Man indeed has fallen, but to which restoration is possible through Him Who is the brightness of His Father's glory, and the express image of His Person."

Chapter VI

Theistic Evolution

A "theist" is one who believes in a God, "theism" being simply the opposite of "atheism." A Mohammedan is a theist. Hence "Theistic Evolution" signifies the acceptance of the theory in a form which does not deny the existence of God. As a matter of fact, the term "Theistic Evolution" is little more than a name. Those who have brought forward and have popularized the doctrine of Evolution are not in the least concerned about "Theism." Their aim has ever been to abolish God altogether, or at least (since a "First Cause" is essential to the theory) to deprive Him of all personality and attributes, and to banish Him to the remotest confines of time and space.

Much less are evolutionists concerned about Christianity, except to antagonize its vital truths. Evolution was put forth as an anti-Christian and infidel doctrine; and for fifty years it has supplied the platform from which and the weapons with which, Christianity has been assailed. Haeckel, the infidel naturalist, termed Darwin's Origin of Species the "Anti-Genesis," and exultingly proclaimed that "With a single stroke Darwin has annihilated the dogma of creation." This antagonism between Evolution and Christianity is a fact which, we suppose, no sincere evolutionist would deny.

Nevertheless, there has arisen in recent years a large class of theologians who, while choosing to call themselves "Christians," nevertheless accept and advocate the doctrine of Evolution. These have attempted to effect a compromise between the two irreconcilable systems, and to that compromise, they have been pleased to give the name "Theistic Evolution." They would hold to Evolution as a general cosmic process but would put it under the control and supervision of God, and would allow of Divine intervention by direct action at those stages which evolutionists find it particularly hard to get over. They would allow just so much "Theism" as seems necessary to help Evolution over the hard places. But inasmuch as this compromise permits enough Divine action in the affairs of the universe to destroy the theory of Evolution, as set forth by the responsible exponents thereof, we may dismiss "Theistic Evolution" as a mere verbal expression to which there is, and can be, no corresponding reality. True evolutionists would not recognize such a self-contradiction as "Theistic Evolution."

In this connection, we quote further from Prof. Fairhurst:

"The first great evolutionists, beginning with Darwin, and including Huxley, Spencer, Tyndall, and others, based the theory of evolution on matter, motion, and force. It was purely a system of naturalism, that did not recognize God, nor the Bible, nor what the Christian regards specially as the supernatural."

"No cosmic evolutionist can accept a miracle at any point of the natural process. To him, a miracle as a part of Evolution would be unthinkable."

Thomas Huxley speaks quite as plainly as Haeckel, saying: "Not only do I hold it to be proven that the story of the Deluge is pure fiction; but I have no hesitation in affirming the same thing of the story of the Creation."

According to Herbert Spencer, nothing is known of God except that He is "unknowable." If this is not practically the same as denying the existence of God, it would not be easy to say wherein the difference lies. If there be a Supreme Being, and He is "unknowable," then it must be either because He has not the power to make Himself known, or because He has not given to the highest of His creatures the capacity to know Him. The first supposition is disposed of by the consideration that, if God did not have the power to reveal Himself and to create beings capable of knowing Him, He would not be God. And the alternative is disposed of by the fact that Man actually possesses the faculty of reflecting upon God, that he has a consciousness of God, and that he has the ability to understand communications from others equal or superior to himself in the scale of being.

Mr. Spencer dogmatically asserts that "the deepest, widest, and most certain of all facts" is this, namely, "that the Power which the Universe manifests to us is utterly inscrutable" (First Principles, p. 46). This is, for all practical purposes, pure atheism. It asserts that there is no revelation from God and can be none. It is, however, an assertion of the most reckless sort, which has absolutely nothing to back it up except Mr. Spencer's spiritual blindness and deadness. It has no more weight or authority than would attach to the assertion of a blind man that the deepest, widest, and most certain of all facts is that total darkness is the universal and perpetual state of nature. That a man may be in complete ignorance of God is evident enough, but that one should make his own ignorance the ground of denying the possibility of knowing God is simply to add colossal presumption to total ignorance.

It requires no great penetration to see that the real object of attack by the supporters of Evolution is the Bible, with its revelation of Christ as the Redeemer and Savior of men. It matters little or nothing whether a perishing child of Adam believes in the existence of God or not, so long as he is blinded to the one thing which most concerns him to know, and that is the salvation of God, which the Bible reveals, and which is received by all who believe "the testimony that God gave of His Son."

That Evolution serves most effectually to blind the minds of all who accept it to the facts of sin, and Redemption is undeniable. Therefore, the pretense, masked by the term "Theistic Evolution," that the doctrine can be reconciled with the truth of Christianity, is merely an attempt to make it more successfully destructive, by throwing incautious people off their guard. There is not a single deadly heresy, among all that were, in past generations, openly opposed to the faith of Jesus Christ, which has not now succeeded, by one means or another, in entering into and establishing itself within the precincts of professing Christendom, and which is not, in our day, openly preached and taught in the "churches" and theological seminaries.

When the main features of the present state of Christendom, as briefly outlined above, are understood, there will be, as Prof. Graebner has well said, "no need to inquire why, on the one hand, enemies of the Bible in all ranks of life greeted with such joyous acclaim the principle announced by Darwin, and why, on the other hand, a chief purpose of Christian apologetics has become the demonstration that Christianity is justified even by reason in that view of the origin of the world which it inculcates, and that, on the other hand, the evolutionary hypothesis is contradicted by the facts of religion, of history, and of natural science."

The spread of the doctrine of Evolution has been phenomenal. Therefore, many theologians became alarmed "because they thought that 'Science' had succeeded in proving that all things were produced by Evolution. They began to consider how they could reconcile theology and 'Science.' They imagined that evolution was an established science. They said: 'We will change the lion into a lamb by changing its name.' And so they called it 'Theistic Evolution,' but accepted the agnostic or atheistic method, and then began to sleep comfortably over their wisdom (?)"(Fairhurst).

It is indeed a fact that those "Christians" who have thus surrendered to infidel Evolution have done little more than devise a name.

Evolution and Christianity

Between Evolution and Christianity, there is and can be nothing but the sharpest antagonism. Prof. Fairhurst well says, "Christian evolution is inconceivable."

Christianity is based upon the fact that the Bible is a Divine revelation. But the Bible, according to Evolution, is itself but a detail of the cosmic process. Here is an issue as to which reconciliation is impossible. One cannot hold Evolution and hold the Christian view--which is Christ's own view--of the Holy Scriptures. If the Bible is from God, if every Scripture is God-breathed, if holy men of old spoke as they were moved by the Holy Ghost, then Evolution is false, and its author is that father of lies, whose chief aim is to "deceive the nations" and to "blind the minds of them that believe not."

One of the best known writers of our day, Mr. H. G. Wells--himself a thorough-going evolutionist--has lately declared in print that Civilization owes both its origin and its preservation until now, to the Bible, saying: "It is the Book that has held together the fabric of western civilization"; it has "unified and kept together great masses of people"; and in fact "the civilization we possess could not have come into existence and could not have been sustained without it." And Mr. Wells drives his point to its logical conclusion by showing that, without something to effectively take the place of the Bible, civilization will speedily be overthrown.

This frank admission involves, if true, the complete negation of Evolution. For, according to that theory, the Bible should be the product of Civilization, and man's ever-advancing Progress should be continually producing, by slight variations, better and better Bibles. But here is an evolutionist who forgets his doctrine long enough to declare that the Bible produced Civilization, and not Civilization the Bible. Here then, in that ancient Book, which is forever correcting and improving man, but which receives no correction or improvement from man, we have a complete disproof of Evolution. What we here assert is, not merely that the statements contained in the Bible contradict the doctrine of Evolution, but that the very existence and persistence of the Bible, in its place of undisputed supremacy among books (a place it holds despite the most strenuous efforts to dislodge it); the hold it establishes upon the hearts and consciences of men; the stupendous and morally excellent influence it has exerted upon the lives of individuals and the prosperity of communities; constitutes a proof of the most convincing sort that Evolution is a monstrous falsehood. If Evolution were true, the history of the Bible, with its place and influence among men, would have been an impossibility. Hence the existence of the Bible is a disproof of Evolution.

The Law and the Gospel Not Evolved

The law of Moses, with the peculiar economy based thereon, and the peculiar people associated therewith--the Israelites--were not the product of Evolution. The children of Israel came out of Egypt utterly unorganized, having lived there for centuries in slavery, dominated by an idolatrous and polytheistic race. At the time of their departure from Egypt, they had neither laws, government, worship, nor organization. Yet they entered Canaan forty years later with a law, statutes and judgments, and a system of monotheistic worship, utterly unlike anything previously existing in the world. The miracles recorded in the books of Moses explain what otherwise would be inexplicable. Judaism is a complete refutation of the theory we are discussing.

But if the Jews, and their laws, institutions, and worship cannot be accounted for by Evolution, still more impossible is it to account for Christians and Christianity by that theory. Christianity was not the product of Evolution. There were no "resident forces" in the world leading gradually up to it; no progress towards it; but just the reverse; for everything was going rapidly in the opposite direction. Judaism had departed completely from the spirit and teaching of the law and the prophets. Greek advancement in literature, philosophy, and art had eventuated in a puerile system of polytheism, and in extreme moral degradation; while Roman progress in the art of government had produced atheism and unspeakable corruption and decay in morals. Christianity arose, not only utterly different in every feature from its environment, but in deadly antagonism to the tenets of Jews, Greeks, and Romans. Christianity considered merely as a historic fact, in connection with its environment, destroys Evolution down to the ground. There is but one conceivable explanation of Christianity, and of the people who "were called 'Christians' first at Antioch," and that explanation is Christ; the Christ of the Gospels, born of a virgin; the Word made flesh and dwelling among men, as Immanuel, God-with-us; Christ crucified, and Christ risen from the dead; "Christ the power of God, and the Wisdom of God" (1Co.1:24).

Evolution and Christ

To say that Jesus Christ was evolved, that He was the product of His environment, is both to repudiate Christianity, and also to reject the plainest facts of history.

Here we reach the climax of the matter. Christ is "the Truth"; and the conclusive test of every doctrine and every system is to bring it into the presence of Christ. When subjected to that test, Evolution fades into nothingness like the mists in the presence of the sun. The resurrection of Jesus Christ from the dead was not an evolution. It was a complete reversal of the course of nature. The people who are "quickened together with Christ" are not an evolution, but a "new creation."

Here again, we quote a striking passage from Prof. Graebner.

"We cannot leave this subject without briefly adverting to a great historic fact, indeed the most massive and significant fact in all history, which, in its more remote bearings, not only strikes at the very root of evolutionistic philosophy but at the same time wounds it mortally in all its parts. I refer to the Resurrection of our Lord.

"The resurrection of Jesus Christ is the central fact of our Christian faith; and it is, when rightly understood, the all-sufficient answer to the theory of Evolution.

"Christ's resurrection is a historical fact, fully as much as the defeat of Xerxes at Salamis in 480 B.C., the discovery of America by Columbus in 1492, and the peace of Versailles of 1919 are historical facts, proven by the word and record of contemporary witnesses.

"But, if Christ was raised, then we have proof for the following tenets, all contradicting evolutionary speculation at so many vital points: (1) The existence of a Personal God, Who is concerned with human affairs; (2) The reality of miraculous interference with natural forces; (3) The truth of atonement and redemption; and (4) The inspiration of the Old Testament Scriptures (hence also of the creation account in Genesis). The details of the argument are beyond the scope of this paper; but a little patient study will bring to light the fact that each of these four basic ideas is dovetailed, mortised and anchored so firmly in the fact of Christ's resurrection, that you can get rid of them only by denying that fact. Hence it is, aside from any investigation of proofs of Evolution, clear to the Christian student that there must be some fault either in reason or in the observation that vitiates the whole theory. The resurrection of Christ is a fact, to which the entire history of Christianity bears witness, the tremendous fact in the history of the world. And it stands four-square against a theory which says that there is no personal God, no sin, no redemption; that there are no miracles, no revelation, no inspiration; that there is no absolute religion and no absolute standard of right and wrong."

The supreme disproof of Evolution then is the Risen Christ, and the results which have everywhere followed the preaching of the Risen Christ.

Chapter VII

Estimates of Darwinism

The Darwinian doctrine of Natural Selection has been discarded by Spencer, Huxley, and other leading evolutionists, who thus leave the theory of Evolution, as it were, suspended in midair, without any method whereby it could work. Mr. Darwin himself had serious misgivings as to his theory, and never regarded it as established.

We consider that the abandonment of Natural Selection must logically involve the abandonment of the entire doctrine of Organic Evolution. It is appropriate, therefore, to make a brief reference to the very general repudiation in recent years of the Darwinian concept.

Dr. E. Dennert's book At the Deathbed of Darwinism gives the testimonies of leading scientists, showing that the title given to his book is fully justified.

Prof. Luther T. Townsend has also written on The Collapse of Evolution, giving testimonies of prominent men of science to the same effect.

St. George Mivart (University College, Kensington, England) says: "With regard to the conception as put forward by Mr. Darwin, I cannot truly characterize it except by an epithet which I employ with great reluctance. I weigh my words and have present to my mind the many distinguished naturalists who have accepted the notion, and yet I cannot call it anything but a puerile hypothesis."

Prof. Fleischmann of Erlanger, who once accepted Darwinism, but after further investigation repudiated it, says: "The Darwinian theory of descent has not a single fact to confirm it in the realm of nature. It is not the result of scientific research but is purely the product of the imagination."

Prof. Haeckel, a most extreme evolutionist, bewails the fact that he is left standing almost alone. He says: "Most modern investigators of science have come to the conclusion that the doctrine of Evolution, and particularly Darwinism, is an error, and cannot be maintained." And he gives an impressive list of "bold and talented scientists" who, he admits, have abandoned the theory of Darwin, though at one time they advocated it. This admission by one of the most noted infidel evolutionists is important. A house thus sharply divided against itself cannot stand.

Dr. Gotte has published an instructive history of Darwinism, showing the stages through which, it has passed, from its enthusiastic reception down to its final stage "when its days will evidently soon be numbered."

Edward yon Hartman also shows that Darwinism has passed through four stages and says that the opposition has now "swelled into a great chorus of voices, aiming at the overthrow of the Darwinian theory. In the first decade of the twentieth century, it has become apparent that the days of Darwinism are numbered"; and he gives the names of eminent scientists who are "among its latest opponents."

Prof. Joseph Le Conte, of the University of California, says: "The evidence of geology today is that species seem to come into existence suddenly, and in full perfection, remain substantially unchanged during the term of their existence, and pass away in full perfection. Other species take their places apparently by substitution, not by transmutation."

Dr. Robert Watts says: "The record of the rocks knows nothing of the evolution of a higher form from a lower form. ... Both nature and revelation proclaim it as inviolable law, that like produces like."

Dr. J. B. Warren, of the University of California, said recently: "If the theory of evolution be true, then, during the many thousands of years covered in whole or in part by present human knowledge, there would certainly be known at least a few instances of the evolution of one species from another. No such instance is known."

Prof. Owen declares that "no instance of change of one species into another has ever been recorded by man."

George Ticknor Curtis, in a recent book, Creation or Evolution, says: "The whole doctrine of the development of distinct species out of other species makes demands upon our credulity which is irreconcilable with those principles by which we regulate, or ought to regulate, our acceptance of any new matter of belief."

Prof. Dana, in his Manual of Geology, says: "Science has no explanation of the origin of life. The living organism, instead of being a product of physical forces, controls those forces for its higher forms, functions, and purposes. Its introduction was the grandest event in the world's early history."

Lord Kelvin, the very foremost of English scientists in his day, in an address delivered in 1903, said: "Forty years ago I asked Liebig, walking somewhere in the country if he believed that the grass and flowers which we saw around us grew by mere chance force. He answered, 'No; no more than I believe that a book of botany could grow by mere chemical force. ... It is not in dead matter that men live and move and have their being; but in a creative and directive Power, which science compels us to accept as an article of faith. Is there anything so absurd as to believe that a number of atoms, by falling together of their own accord, could make a crystal, a microbe, or a living animal?'"

There is nothing so reasonable as faith. For faith is simply the acceptance of the testimony of God, given to men in "the Scriptures of truth," which have proved themselves, in their history and influence in the world, to be super-human. Conversely, there is nothing more unreasonable than unbelief; for it denies not only the light of Divine revelation but that of nature (Rom.1:20; Act.14:15-17; Act.17:24-29). Small wonder is it that men who vaunt the human intellect and who have rejected the word of the Lord should be so fatuously credulous as to attribute designing skill and creative power to a mere concourse of atoms.

The same Lord Kelvin, whom we have just quoted, is on record as declaring that, there is not a single ascertained fact of science which conflicts with any statement of the Bible.

When, therefore, we hear, as is common enough nowadays, assertions made by unbelieving theologians and others, to the effect that "science" has shown this or that statement of Scripture to be erroneous, let it be remembered that we can bring the testimony of the most eminent men of science to prove those assertions false.

Dr. Ethridge of the British Museum, a noted expert in fossil-ology, speaking of the views of evolutionists, says: "This Museum is full of proofs of the utter falsity of their views."

Prof. L. S. Beal, acknowledged to be in the front rank of British scientists, in an address delivered June 1903, said: "The idea of any relation between the non-living, by gradual advance of lifeless matter to the lowest forms of life, and so onward to the higher and more complex, has not the slightest evidence from any facts of any section of living nature of which anything is known."

Virchow of Berlin, regarded by some as the foremost chemist of the world, said, "It is all nonsense. It cannot be proved by science that man descended from the ape or from any other animal." He went so far as to denounce the theory as dangerous to the state and demanded that it be excluded from the schools.

Much more of the same sort might be added; but it will suffice to refer to Prof. Fairhurst's Theistic Evolution (Standard Publishing Co., Cincinnati), and Graebner's Evolution, already referred to, from which most of the above quotations are taken. We will only mention additionally a statement made in a very recent address (February 1922) by Prof. Wm. Bateson, the distinguished English biologist, a scientist of the first rank, who, speaking in Toronto, Canada, is reported to have said: "It is impossible for scientists longer to agree with Darwin's theory of the origin of the species. No explanation whatever has been offered to account for the fact that, after forty years, no evidence has been discovered to verify his genesis of species."

Surely our "liberal" theologians, who teach as truth that monstrous fiction which true men of science never regarded as more than a speculative theory, and now have, with practical unanimity, repudiated, are utterly without excuse.

The Existing Danger

Notwithstanding the fact that Darwinism is no longer believed in the circles in which it originated, its influence for harm was never so great as now. The reason is that the theory has found its way into the theological seminaries, and into the schoolbooks of the children, where it is doing the deadly and truly devilish work of discrediting, in the eyes of many, the statements of the Word of God.

Darwinism in the Schools

A parent, writing to a religious periodical, tells of a textbook brought home by his seven-year-old boy, the title of which was "Home Geography for Primary Grades." The following quotation will serve to show what is now being taught to children of the most tender years. Discussing the subject of birds, this text-book for primary grades says: "Ever so long ago their grandfathers were not birds at all. Then they could not fly, for they had neither wings nor feathers. These grandfathers of our birds had four legs, a long tail, and jaws with teeth. After a time, feathers grew on their bodies, and their front legs became changed for flying. These were strange-looking creatures. There are none living like them now." Such is the monstrous fictions now taught to little children as scientific truth.

It is a significant and disquieting fact that a determined effort recently made in the legislature of Kentucky to forbid the teaching of Evolution in the schools of that State was defeated. Thus the arch enemy of God and men have maneuvered this "Christian" country into the position where the accepted canons of education forbid the teaching of the Bible to the children of the tax-payers, but permit the teaching of the most anti-Christian and unscientific doctrine that ever made a bid for public favor.

The support for the teaching of Evolution (which in practically all cases means the utterly discredited theory of Charles Darwin) came from "educators and religious leaders" (so says the Literary Digest, March 25, 1922) like Lyman Abbott, Dr. Angell, President of Yale, Dr. Lowell, President of Harvard, and Dr. McFarland, Sec'y of the Federal Council of Churches.

It is high time for parents to be awakened out of sleep as to the dangers to which their children are exposed in our modern schools. These are indeed "perilous times"; and one of the greatest perils thereof is the teaching which is now being given to the young. Parents, who would be careful to keep their little ones from the dangers of the streets, recklessly expose them every day to the more serious dangers of the schools and give themselves little concern as to what they learn there from teachers and companions.

Chapter VIII

Evolution in Human Affairs

We come now to a matter of deepest interest and importance; for when we turn our attention to the realm of human affairs and activities, we observe a state of things which is in the greatest possible contrast to what is seen elsewhere among living species. It is exceedingly important, with a view to a right understanding of the theory of Evolution, that this contrast is noted, and its significance is comprehended.

Briefly stated the contrast lies in this, that Evolution is the method of working which prevails everywhere, and always has, in human affairs, whereas outside of human affairs there is not a trace of it to be found in all the universe. By "human affairs," we mean, those activities wherein man himself is the designer and agent. For there is a realm wherein man is the directing and controlling authority, wherein he has free scope to try out all his ideas, and to exert all his powers in every direction.

Man, in all his operations, and in every department of his diversified activities, is progressive. Other living creatures are absolutely unprogressive. Man develops arts, industries, social institutions, governments, etc., etc., by trying experiments, discovering defects and weaknesses, devising remedies, and so on, the changes being so rapid and so extensive that each generation lives in a different industrial, political, social and religious environment, to that of its predecessors.

In Society at large, we find a typical illustration. It is one employed by Herbert Spencer. He says (First Principles ch. 14. Sec. 3) "In the social organism integrative changes are clearly and abundantly exemplified." And so beyond dispute they are. Mr. Spencer cites the development of "Society," beginning with wandering families, then tribes, then stronger tribes formed by union with or subjugation of others, until the combinations became relatively permanent, and ultimately were evolved into States and Nationalities. Nor has this progression ceased; for since Spencer's time there have been further combinations of nations, and finally a "League of Nations," which will undoubtedly eventuate in the Federation of Kingdoms, symbolically pictured as the Beast in Daniel and Revelation.

The same progression from simple and incoherent beginnings, to conditions relatively complex and coherent, may be traced in every department of human affairs. Whether we examine the industrial groups, the ecclesiastical, the military, the medical, the legal, etc., we find the same progressive development.

Let us consider a few illustrations of this striking law of humanity.

A few centuries ago the crudest implements served the farmer for preparing the soil and gathering his crops. From those simple beginnings have evolved the tractors, harvesters, and other modern wonders of farm-equipment; and the advance has been by slight, progressive changes. Here is Evolution sure enough, and precisely as described by Spencer and other materialists.

So likewise, in the department of Locomotion and Transportation, it is easy to trace, between the wheel-barrow and ox-cart of by-gone days, and the auto-car and flying machine of the twentieth century, a connected line of evolutionary progress. And a similar line may be traced from the birch-bark canoe to the Trans-oceanic liner and the submarine.

If we look along other industrial lines, as milling, printing, paper making, communicating intelligence to distant points, weapons of war, etc., etc., we see the same thing, that is to say, a very crude and imperfect beginning, with a succession of forms, each an improvement upon its predecessors, and with never an end to the development.

Again, in the literary field, we can readily trace the activity of man from a simple beginning in oral recitation and manuscript copies on vellum or papyrus, to the manifold present-day output of books, newspapers and other periodicals.

So, with the religions of the world. The human element in these has undergone great and progressive changes, both in ideals and in forms and observances; and the progress continues. Here we have again an instructive contrast; for we can readily compare the development of the religions of the world with the progressive Revelation of the Truth of God. The former follows, like everything else which is under the control of man, a strictly evolutionary course, every new stage involving the destruction of what preceded. The latter is, like everything that comes from God, perfect (as far as it goes) from the start. And, though His Revelation has been given at sundry times and in many distinct parts, yet there is not a trace of Evolution in it; for every part of God's Revelation remains forever true; and all the parts together unite in perfect agreement to constitute a complete and harmonious system of Truth.

It was a very easy matter to multiply our illustrations, for they lie all around us in plain view. For wherever we look within the realm of human affairs, the pieces of evidence of Evolution stare us in the face. But, in striking and significant contrast with this is the fact that the moment we pass the boundaries of that realm, we strain our eyes in vain for a scrap of evidence to indicate that the process of Evolution ever had a foothold. The birds construct their nests, the beavers their dams, the beehives, and ant-colonies carry on their complex operations, precisely as they always have done. Moreover, each of those creatures does its work perfectly at the very first attempt, whereas man makes innumerable failures before he can do anything even passably well.

We do not mean to say that the law of development or progress, which is seen everywhere in continuous operation in human affairs, corresponds in all respects to the "Evolution" of Darwin, Spencer, and Huxley; for there are great differences. For example, the progressive changes in human institutions are not brought about by blind "resident forces," but by the intelligent actions of men, and with a definite purpose in view. Nor is there a direct organic connection between the improved forms that are constantly springing up and the older forms which they displace. Nevertheless, the resemblance is sufficiently close on the surface, to make it possible for the evolutionist to draw his illustrations from the wide field of human activities and that is the important matter we wish to press upon the attention of our readers in this chapter; for without the help of those illustrations the theory of Evolution could not possibly stand.

Evidences of Evolution in Human Affairs

In this connection, there are several facts which have an important bearing on the main question, and which should, therefore, be carefully noted.

First, in the field where Evolution does operate--that is, in the realm of human enterprises, from which Spencer and others draw all their illustrations--the evidences of its workings abound. Traces of the earlier and cruder forms, which subsequent improvements have displaced, are plentiful. The advancement, moreover, is not so slow as to give the impression that things are at a standstill, but, on the contrary, is sufficiently rapid to permit of observation of its character and direction. From these facts, it must be concluded that, if there were any Evolution in those realms of nature which are not under the guidance and control of "the will of man," there would be abundant pieces of evidence of its workings in those spheres also. The only and the sufficient reason why things in Nature appear to be at a standstill and have so appeared during the thousands of years they have been under man's observation is that they are at a standstill. The simple and satisfactory explanation of the fact that no trace of Evolution has ever been found in Nature is that there has been no Evolution there.

Second, it is seen that in all the departments of human activities there is never any end of development, either in the construction of the things which man makes or in the methods by which he operates. Never, in any part of this realm, is a stage reached where there are rest and stability. Never has the right thing, or the right method, been attained. But in Nature all is stable. Both structures and processes remain identically the same as they have ever been.

Why this astonishing difference? Manifestly, the reason why there is no improvement in the life habits of the lower orders is that there is no need of any. Indeed, we can say there is no possibility of any. For who could improve upon the structure or materials of the honeycomb? The ways of those creatures do not change, for the simple reason that their ways, works, and habits of life are just what their Creator planned for them.

With man, it is far otherwise. The scheme of life appointed to him has been completely wrecked. His whole race is blighted and degraded. It finds itself in conditions of poverty and wretchedness. Its energies, therefore, are directed towards improvement, for the very reason that it has fallen from the place of life appointed for it.

But the most learned and intelligent of those who reject the light of Holy Scripture (which reveals the truth as to man's condition, and shows that, despite all material gains, corruption and decay still increase and spread among the children of men) are easily misled as to the facts; and they mistake material gains for true progress.

Man's "progress," of which he loudly boasts, is a delusion. There is indeed a constant advance in mechanical inventions, and in all that contributes to grossly materialistic industrialism. But that apparent progress serves but to hide the real facts from the many, who look only at the surface of things. In reality, the civilized nations are on the downgrade. Crime, insanity, and suicides increase at an appalling rate. Everywhere the spirit of violence and lawlessness is rampant. In commercial life, political life and private life corruption increases and abounds.

The noted scientist, A. R. Wallace, who put forth the theory of Evolution contemporaneously with Darwin, said, shortly before his death in his ninety-first year:

"I have come to the general conclusion that there has been no advance either in intellect or morals since the days of the earliest Egyptians. Everything is as bad as it possibly can be. There exist in our midst horrors and dreadful diseases never known before. Our whole social environment is rotten, full of vice, and everything that is bad."

Had Mr. Wallace lived to see the conditions in the world resulting from the great European War, he would have had to admit that things could be even worse than they were then.

The Spread of the Theory Accounted For

Finally we call attention to the pertinent fact that the presence and working of the law of Evolution in human affairs has furnished Spencer and others with such apt illustrations, and they have used them so skillfully, that many fail to see the important fact to which we are here calling attention, namely, that the instances of Evolution to which they are able to point lie always in the realm of human institutions.

To this point we ask careful attention, for in it is found the explanation of what every intelligent person will ask, namely, how comes it that a theory, for which there has never been the slightest proof, but which, on the contrary, is opposed to all the facts of human observation, has met with such wide acceptance among intelligent people? The reason is that, in the field which lies nearest to man, and in which he can most easily observe, there is Evolution; not only in localities but everywhere; and not only occasionally but always. It was easy, therefore, to make the assumption that, in the more remote fields of Nature, the same law of progressive changes was in control; and the fact that "varieties" of plants and animals could be produced by artificial breeding, gave a color of support to the theory. That theory once formulated and proclaimed, it would inevitably be received exultantly by all who are biased against the truth of man's fall and depravity, and so it was sure of strong and enthusiastic support.

The Fatal Bias

On this point we quote again from Prof. Graebner:

"The warfare of philosophy against Christian faith is readily explained. Man is corrupt. He loves sin. He is conscious of his guilt and fears the penalty. Hence every avenue of escape is welcome if only he can persuade himself that there is no God, no judgment. Man is proud; he desires no Savior. Hence the effort to prove that no Savior is needed, that there is no guilt attaching to sin, that there is no absolute right and wrong. Hence too the doctrine of the Agnostic that we can ascribe no attributes to God. When we read the 'Synthetic Philosophy' of Spencer we are apt to believe that the agnosticism there set forth is the result of deep philosophical speculation. Nothing could be further from the truth. Man, even cultured philosophical man, wants to have no restrictions placed upon pride and selfishness; hence it is necessary to rid the mind of the fear of Divine justice; hence the desire to demonstrate that God has no attributes, such as that He is 'just,' for instance. The Psalmist describes this attitude in the words, 'Let us break their bands asunder, and cast away their cords from us.'

"No one who has grasped the inner motive of all 'Scientific' effort to demolish faith, can fail to understand why the many greet with such jubilant acclaim every new attack upon the Biblical narrative. No one who has pondered this motive can be snared in the net of 'science, falsely so-called.' He has seen its inwardness; he knows its fatal bias."

The Law of Sin and Death

The law of Evolution then is strictly confined to one realm, the sphere of human affairs; and therein it has full sway. It is the law or rule of action of a fallen race. It is "the law of sin and death" (Rom.8:2). It has no place in the Kingdom of God, or in any sphere which sin has not invaded. It arises solely from man's efforts to improve his wretched condition, and from blindness to God's way of recovering and restoring His perishing human creatures.

Man, having discovered that Evolution is the rule of procedure in the realm of his own doings, has imputed the same law to his Maker, thus fulfilling the Scripture, "Thou thoughtest that I was altogether such an one as thyself."

FALLACIES OF EVOLUTION: A SERIES OF THREE LECTURES

by John D. Charles

Introduction

Great difficulty is being experienced in finding textbooks in science that are free from the unscriptural teaching of evolution, or the descent of the species. This is especially true of textbooks on biological subjects. The baneful effect on the minds and lives of students who receive their training in science from such textbooks is very apparent. That there is no conflict between true science and the Scriptural account of the Creation and the condition and development of both animal and vegetable life is conceded by many of the best scientists and theologians—in fact, it is the only position allowable to those who accept the inspiration and inerrancy of the Word of God.

The conviction that the Bible account can be fully established by scientific facts and the great need of teaching the harmony between the Bible record and scientific investigation gave rise to the contents of this little volume.

These three lectures were given before the student body during the Special Bible Term of Hesston Academy and Bible School in the month of January 1916 and published in the Gospel Herald soon after. Calls for the lectures in permanent form came from various sources, both educational and theological, from the home and the institutions of learning, and the author was persuaded to arrange his work for publication in book form.

The treatise has a place as a reference work for students in both scientific and theological study. Its bibliography alone is of great value. Moreover, it provides interesting and instructive reading for the home and the library. Its mission is to establish and confirm the student in the faith of the truth of the Bible along the line that it is most frequently attacked by scholars and ripened intellectuality, and at the same time give correct information on questions of science relative to life and its development. May the perusal of its pages and the study of its teachings tend to magnify Him who originates, designs and controls the destiny of all life mortal and immortal.

D. H. Bender.

Lecture I—The Question at Issue

In nearly every field of activity, for the last half-century writers and speakers have been illustrating or endeavoring to prove their discussions by appealing to evolution as popularly believed with reference to things animate and inanimate, material and immaterial. Evolution is claimed by its advocates to be the greatest intellectual discovery of the nineteenth century, and by some, especially followers of the theory, the greatest thought that ever entered the mind of man. There is possibly no subject on which writers in these days become so easily eloquent as the universality of the great theory of evolution, yet there is no subject which gives a louder call for careful discriminations and cautious statements.

Definitions

The word evolution is derived from the Latin evolvere which means to unroll. In a perfectly legitimate and inoffensive sense, it has come to be used to denote any process of advancement, growth, or development. For instance, in biology evolution may simply mean the development of an organism. The gradual growth and unfolding of the parts and organs of a plant from the seed is an evolution. The flower is the evolution of the bud. The development of an animal from conception to old age is properly called an evolution. The chicken is the evolution of the egg. Thus, it is seen that the word has a perfectly innocent meaning with a large range of legitimate uses in the very fields where, as we shall see, it comes into much-deserved disrepute.

The Bible, indeed, teaches a system of evolution. Matter, animate and inanimate, as we see it in the world about us, was not made in a moment or even in one day, but in six days. In Genesis, we find that the whole process of creation was an orderly progress from lower to higher forms of matter and life. The Bible account of the creation reveals an established order in all the Creator's work. So everywhere, in common use of the term, there is evolution. There is "first the blade, then the ear, then the full corn in the ear." Even the Kingdom of Heaven is like a mustard seed, "which a man took, and cast into his garden; and it grew, and waxed a great tree; and the fowls of the air lodged in the branches of it."

But during the last fifty years the word, in another sense, has come into much-merited dishonor. Just as we speak of the evolution of the seed into the plant and the still further evolution of the bud, flower, and fruit, meaning by our words the natural onward action of the life within, so certain evolutionists apply this same principle to the whole universe which they say came in the same way. This idea may be illustrated by quoting from W. H. Norton, professor of geology in Cornell College. He says: "Just as growth is characteristic of the individual life, so gradual, progressive change, or evolution, has characterized the history of life upon the planet. The evolution of the organic kingdom from its primitive germinal forms to the complex and highly organized fauna-flora of to-day may be compared to the growth of some noble oak as it rises from the acorn, spreading loftier and more widely extended branches as it grows." (Elements of Geology, page 301).

The meaning of the word is thus extended, in modern usage, not only to the development of an individual organism, but is also supposed of a race, a species, or other group. And here is where the trouble arises. As I have written elsewhere, "The term evolution need not in itself alarm the Christian. We have no objection to it if understood to mean the fulfillment of the plans of the Creator in nature as to the individual growth, structure, function, and variation of the species. The kind of evolution which we do object is the spontaneous evolution of nothing into matter and energy, and of these into all the wonderful and complicated forms of growth and of the origin of species, which it considers yet to be in a condition of change to something higher." (Bible Doctrine, page 60.) It is only by the injection of erroneous and harmful philosophical and theological implications into the term evolution that it becomes disreputable.

To grasp in a general way the popular theory of evolution let us notice a few definitions by scientists and others who are qualified to speak. In gathering these statements, I was impressed with the fact that evolution, that magic word which many people fancy explains everything, is itself understood perfectly by none. At least the first definition I shall give that of Herbert Spencer will help but a few of us to understand the theory. In his "Data of Ethics," page 65, he explains evolution as, "A change from an indefinite coherent heterogeneity to a definite coherent heterogeneity through continuous differentiations and integrations." I shall not take time here to try to explain this as whatever ideas may be contained will doubtless be brought out in the following definitions.

Possibly the best definition to give us an intelligent idea of the theory at once is the one framed by Dr. L. T. Townsend in his "Collapse of Evolution" page 8: "Evolution in its biological restriction, involves the theory that living things originally came upon the earth in the form of germs, supernaturally created, or imported, or produced by spontaneous generation, and then through natural and orderly processes, long-continued, developed into various species of plants and animals, existing and extinct, culminating in man, who is recognized as the crown and glory of all earthly things."

A professor in Wesleyan University states the case thus: "When we trace a continuous evolution from the nebula to the dawn of life and again a continuous evolution from the dawn of life to the varied fauna and flora of today, crowned with glory in the appearance of man himself, we can hardly fail to accept the suggestion that the transition from the lifeless to the living was itself a process of evolution."

Lyman Abbott tells us: "Evolution is a doctrine that this life of man, this moral, this ethical, this spiritual nature has been developed by natural processes."

Tyndal's statement of the scope of the theory is as follows: "Strip it naked and you stand face to face with the notion, that not only the ignoble forms of life, not only the more noble forms of the horse and lion, not only the exquisite mechanism of the human body but the human mind with its emotions, intellect, will and all their phenomena, were latent in that fiery cloud." (Christianity and Positivism, page 30.)

Weismann, in his last work, said: "There can be no question that man has evolved from animal ancestors, whose nearest relatives were the anthropoid apes."

From Darwin's "Descent of Man," page 372, we read: "Man is descended from a hairy quadruped furnished with a tail and pointed ears, probably arboreal in its habits and an inhabitant of the Old World. This creature, if its whole structure had been examined by a naturalist, would have been classed among the Quadrumana as surely as would the common and still more ancient progenitor of the Old and New World monkeys. The Quadrumana and all the higher mammals are probably derived from an ancient marsupial animal, and this through a long line of diversified forms, either from some reptile-like, or some amphibian-like creature, and this again from some fish-like animal. In the dim obscurity of the past, we can see that the early progenitor of the Vertebrata must have been an aquatic animal provided with branchia, with the two sexes united in the same individual."

Should any say that Darwin's theory is no longer held, here is a modern version of the same ideas. A. S. Packard, in his discussion of evolution in the New International Encyclopedia, issued in 1911, says: "Life appears to have been a necessary and inevitable result of inorganic or cosmic evolution. It came into being on our planet in the most natural way.... It arose by the action of physicochemical laws through what we call spontaneous generation, the materials for the first bit of living photo plasm being ready at hand.... Man's origin from some mammal is strongly attested by the presence in his body of a large number of vestigial characters, which indicate an ancestor that went on all fours—some features appearing shortly before and after birth, hinting at an ape ancestry."

It is not necessary to extend these expressions. No one is ignorant of the theory today for it is being taught in high schools, academies, colleges, universities, and theological seminaries. The theories of evolution as we have noticed from these expressions vary in some details, but all agree in saying that man was descended from the brute as to his body at least and some even claim this descent for the whole man. This teaching as to man is the highest and most vital part of the whole theory, and in this all popular evolutionists are practically agreed. So that as far as their effect on Christian doctrine and God's Word is concerned, all may be classed together. As George Frederick Wright, an eminent geologist of Oberlin College, says: "The widely current doctrine of evolution which we are now compelled to combat is one which practically eliminates God from the whole creative process, and relegates mankind to the tender mercies of a mechanical universe the wheels of whose machinery are left to move on without any immediate divine direction."

We now turn to a brief history of evolution. How did such a stately theory as is held today come about? It is not a new theory. Aristotle, who lived B. C. 384–322, may be regarded as the father of the theory of descent. This idea in crude form was held by ancient philosophers, church fathers, and scientists from a very early date.

But the true founder of these ideas was Lamark (1744–1829). In 1801 he published his views. They were not generally accepted but misrepresented or ignored through the influence of Cuvier, the French scientist. It may be noted in passing that the French people have never taken to these evolutionary ideas even though Lamark was a Frenchman.

It was reserved, however, to Charles Darwin, in 1859, and seconded by A. R. Wallace, who died recently, to convert the majority of the scientific world to the popular evolutional views. These men separately announced the hypothesis of (to use Darwin's own words) "the origin of species by spontaneous variations, and the survival of the fittest through natural selection, in the struggle for existence." At first scientists with some hesitancy adopted these views, but later they were so generally accepted in Germany, England, and America that by about the year 1885, anyone who did not fall in line with the theory was branded unphilosophical, unscientific and unscholarly.

As early as 1852, Herbert Spencer in England advocated evolution from a philosophical point of view. He used the term both for the inorganic and organic world and even extended the doctrine of descent to history, society, morals, ethics, and religion. Since his day much evolution has been in the world.

Indictment

I have already intimated heavy indictment against this now popular theory. I wish to enter more fully into the charges after which I will give the verdict of scholars of the past and today to substantiate the indictment.

Evolution has since Darwin's time become invested with omnipotence by its most enthusiastic followers. What evolution cannot accomplish and account for is not worthwhile. As James Orr, professor of apologetics and systematic theology in United Free Church. Glasgow has said: "It receives extensions and applications which carry it far beyond the bounds of established fact. From the organic world, it is extended to the inorganic; from our planet and solar system to the cosmos; from nature to the creations of man's mind."

The most serious charge against the theory, however, is that it undertakes to account for the Bible, for what has popularly come to be known as "higher criticism" is the theological brand of evolution. It attempts to reconstruct the Bible. The miraculous narrative is relegated to the realm of mythology. The most sacred doctrines of evangelical faith are discarded. By its claim for the origin of man it vitally affects the sacred doctrines of the nature of man, of sin, and the penalty for sin, man's need, and the work of Christ. According to this theory, Christ Himself, as a mere man, is an evolution. In its radical and most consistent form it is atheistic, and it is a fact that its most ardent supporters were atheists. It is true that many evolutionists of a latter-day are theistic. But it is not enough to believe that there is a God in this matter. The devil believes this. Tom Paine was a theist. But Christianity is far more than theism, that is, simply accepting the existence of a God.

In these lectures, I shall show that, by present-day and true scientific inquiry, this grand scheme planned by scientists, philosophers, men of literature and theologians, held in highest repute, is not well-founded, and unsupported by a single established fact in the whole scope of science, philosophy or religion. Evolution is not accepted by all scientists and much less by all scholars; it is admittedly an unproven theory; it is by its own principles unscientific, unphilosophical, unchristian, and simply a rehash of the old teachings of chance garbed in scientific terms. No single case, in the theoretical and popular use of the term, has ever been shown, and there is no known cause which actuates it. Finally, it will be shown that it is flatly in opposition to the teachings of the Bible and destructive of all Christian faith; that it had its beginnings in heathenism and takes its ending in atheism.

These are serious charges. Unless firmly established by facts that cannot be controverted, these charges must properly be branded as a work of impudence and contemptuous arrogance. But we have every reason to take courage, for we find on our side the verdict of scholarship, to begin with, and the Word of God to sanction it all along the way. By scholarship is here meant thorough literary or scientific attainments.

Verdict of Scholarship

I have in my library a new book entitled, "The Present Relations of Science and Religion," by T. J. Bonney of London, in which I read: "We seem justified in concluding that evolution, as opposed to what is commonly called the creation of species, may now be regarded an established theory." On the same shelf is another book entitled "The End of Darwinism," by Dr. Alfred P. Schultz, in which the author begins his discussion with these words: "The theory of evolution is not accepted by every competent mind, and it is not accepted for compelling reasons. The truth is, that not one fact has been discovered which substantiates the theory, and there are many facts which prove it to be erroneous, not only erroneous but impossible." After carefully searching through a great number of works, pro, and con, I can, even putting aside all Biblical argument, unhesitatingly take my stand with Dr. Schultz and his like for the weightiest testimonies are all on his side.

This is true in spite of the fact that it is repeatedly stated by evolutionists that all scholarly men, scientists, philosophers, and theologians are on their side and that those who question the theory are incompetent and antiquated. More than likely it is possible for this brand of professors to be antiquated because they hide in their own interests and are either unwilling or unable to generalize and apply true scientific methods of instruction.

Here I wish to throw out a warning. We hear the cry far and near that scientific men have abandoned Darwinism. This is in part true, but it is not to be taken that the whole fallacious theory of evolution is abandoned. To quote from one of our ablest preachers: "It is quite fashionable for scientific preachers to whack Darwinism as an exploded theory, without saying a word as to the correctness or otherwise of the present popular theory of evolution. At other times the same fellows who whack Darwinism talk as if anyone who testifies against the present popular theory of evolution is a fool." I heartily agree with this statement. Let not orthodox people rejoice as if the victory were won when they hear that Darwinism has been exploded but remember that these very scientists making this cry have a popular evolution theory which after all is founded on Darwinism and is no better than the one proposed by Darwin. As L. T. Townsend puts the question: "Is it not rather ungracious for evolutionists to kick and desert Mr. Darwin, who spent a lifetime of thought and work in their belief, especially when they are unable to offer any substitute for his abandoned theory of descent?"

I have in my library a number of books that do this very thing, and whenever I read them, I feel that the authors are performing a cowardly trick. They are hiding their own fallacious theory behind the master to whom after all, they are indebted for their own ideas. Poor Darwin! One of these books is a splendid refutation of Darwinism, but in conclusion, I find a statement favoring the theory of descent. Let us be on our guard here and remember that anything that leaves out God as a working factor and admits the organic connections, the survival of the fittest and the natural selection of Darwin, and the liberal claims of Spencer is in every sense profanely evolutionary.

Coming back to the idea that "the consensus of scholarship" is on the side of evolution, I am forced to dissent even to the extent of saying that the case is entirely reversed. Here I find myself in company with Townsend, Patterson, and others who have written on this subject. I now offer quotations from eminent scientists and philosophers who do not support the present popular theory of evolution.

Here are a few statements from George Frederick Wright from whom I have already quoted. Speaking of evolution he says: "It is a fad of the present, which is making such havoc and confusion in the thought of the age, leading so many into intellectual positions, whose conclusions they dare not face and cannot flank, and from which they cannot retreat except through the valley of humiliation." (Bibliotheca Sacra, April 1900.) In another article, he says: "Doubtless the passing of the present phase of evolution is not final. Theories of evolution have chased each other off the field in rapid succession for thousands of years. Evolution is not a new thing in philosophy, and such is the frailty of human nature that it is not likely to disappear suddenly from among men. The craze of the last half-century is little more than a recrudescence of a philosophy which has divided the opinion of men from the earliest ages. In both Egyptian and East India mythology, the world and all things in it were evolved from an egg; and so in the Polynesian myths. But the Polynesian had to have a bird to lay the egg, and the Egyptian and the Brahman had to have some sort of a deity to create theirs.... Modern evolutionary speculations have not made much real progress over those of the ancients. As already remarked, they are in their bolder forms atheistic; while in their milder forms they are 'deistic'—admitting, indeed, the agency of God at the beginning but nowhere else." (The Triumphs of Truth, page 165.) Prof. Wright in his introduction to "The Other Side of Evolution," by Alexander Patterson, recently revised, begins: "The doctrine of evolution as it is now becoming current in popular literature is one-tenth bad science and nine-tenths bad philosophy.... Mr. Patterson does well to emphasize the fact that orderly succession does not necessarily imply evolution from resident forces." In a personal letter to me a few years ago Prof. Wright lamented the fact that it was impossible to obtain science texts free from objectionable and false teaching on evolution.

Dr. Virchow (1821–1902) German pathologist and founder of the germ theory opposed evolution. He was at one time called "the foremost chemist on the globe." On one occasion he said: "The reserve which most naturalists impose on themselves is supported by the small actual proofs of Darwin's theory. Facts seem to teach the invariability of the human and the animal species." At another time he said: "It is all nonsense. It cannot be proved by science that man descended from the ape or any other animal.... All real scientific knowledge has proceeded in the opposite direction." (Lecture: Freedom of Science.) And again, he said: "The attempt to find the transition from animal to man has ended in total failure. The middle link has not been found and never will be."

George Howison, in his book, "Limits of Evolution," writes: "Under the sheen of the evolutionary account of man, the world of real persons, the world of individual responsibility, disappears; with it disappears the personality of God."

E. B. Wilson, professor of zoology, Columbia University, and authority on the cell in this country, writes: "The study of the cell has seemed to widen rather than to narrow the enormous gap that separates even the lowest forms of life from the inorganic world." (The Cell in Development and Inheritance, page 434.) Since the publishing of the book from which this quotation is taken, I have been privileged to take Dr. Wilson's lectures in cellular biology and was more than once impressed with the fairness with which he met all the difficult problems of so critical a study. On one occasion, in referring to a certain theory of Weisman, he said: "If this is true, and I incline to think it is, the whole theory of evolution falls."

H. W. Conn, professor of biology in Wesleyan University, has said: "Nothing has been positively proved as to the question at issue. From its very nature, evolution is beyond proof.... The difficulties to an unhesitating acceptance of evolution are very great and have not grown less since the appearance of Darwin's 'Origin of Species,' but greater." (Evolution of Today, pages 107, 203.)

N. S. Shaler, professor of geology at Harvard University makes this statement: "It begins to be evident to naturalists that the Darwinian hypothesis is still essentially unverified. Notwithstanding the evidence derived from the study of animals and plants under domestication, it is not yet proved that a single species had been established solely or mainly, by the operation of natural selection." (International Quarterly, Dec.–March 1912–1913).

From Alexander Patterson's "The Other Side of Evolution," I copy the following: "The late Dr. J. H. W. Stuckenberg, of Cambridge, wrote me, that many of the scientists of Germany reject the extreme views of evolution and the inferences which men like Prof. Haeckel, of Jena, have drawn from Darwinism. He quotes Dr. W. Haecke, a zoologist of Jena, the home of Prof. Haeckel, as saying: 'We the younger men must free ourselves from the Darwinian dogma, in which respect quite a number of us have been quite successful.' Prof. Paulsen of Berlin has exposed some of Haeckel's fallacies and regards his reasoning as 'a disgrace to Germany.' He said the mechanical theory for which Darwinism was held to stand, is rejected by such scientists as Naegeli, Koelliker, M. Wagner, Snell, Fovel, Bunge, the physiological chemist, A. Brown, Hoffman and Askernazy, botanists; Oswald Heer, the Geologist, and Otto Hamann, the zoologist. Of Carl Ernst von Baer, the eminent zoologist, and anthropologist, Haecke affirms, that in early years he came near adopting the hypothesis of evolution into his system, but that at a later date he utterly rejected it. The same change occurred in the late Du Bois Reymond and Prof. Virchow, the eminent scientist of the University of Berlin.'" No wonder Prof. Haeckel, the greatest living advocate of evolution, complains bitterly that the scientists of Europe who were once with him, have fallen away.

From Prof. Townsend's excellent little pamphlet, we take the following: "Such men as Louis Agassiz, Joseph Henry, John William Dawson and Arnold Guyot pronounced evolution false and unscientific." And again, "Dr. Etheridge, of the British Museum, one of England's most famous experts in fossil-ology, has passed the following criticism upon evolution: 'In all this great museum there is not a particle of evidence of transmutation of species. Nine-tenths of the talk of evolutionists is sheer nonsense, not founded on observation and wholly unsupported by fact. This museum is full of proofs of the utter falsity of their views.'" I remember a similar expression made by the Curator of the American Museum of Natural History, New York City, when as a member of a class in zoology viewing the bones portending to show the evolution of the horse from a three-four toed animal through thirty-five or forty changes to the present one-hoofed animal.

Prof. Petrie, in "Ancient Egypt," issued 1914, page 192, gives some telling blows against the evolutionary theory, by describing some ancient portraits. M. G. Kyle, Egyptologist, commenting on this in the Sunday School Times, Dec. 25, 1915, says: "These portraits show more clearly than pages of discussion how absolutely futile is any theory of evolutionary development of human life that demands a gradual development and regular progress from the lowest character in the early period to the highest in later times, in Egypt or anywhere else, a theory that purports to find men but little above the brutes, and to trace their evolution up to the highest place among men." He states further that these statues show, "Not a race coming up from the beasts and only attaining full human development at a late period, but of a race made in the image of God in the beginning and debased by sin, now rising higher under good influences, now sinking lower under debasing conditions."

We have hardly begun noting testimony against the theory of evolution and yet sufficient has been said to disprove it. Suffice to notice yet a few expressions from prominent preachers, which naturally will take us to the theory of evolution as it affects other sciences beside biology and geology.

From Franklin Johnston, we have this expression: "The hypothesis is a fallacy.... Our race has made progress under the supernatural revelation, but progress under the influence of supernatural revelation is one thing, and evolution is another."

A. C. Dixon, pastor of Metropolitan Tabernacle, London, speaks on this wise: "Evolution compels us to believe that certain parts of the Bible were written at a later date than is claimed by orthodox believers. It must be remembered, however, that the theory of evolution is a pagan and not a Biblical conception. It is supported by fancy and not by fact. It is imagination run wild. It is a myth born in the brains of scientific men, who have a genius for generalization."

J. J. Reeve, of the South-Western Baptist Theological Seminary, Fort Worth, Texas, gives his experience thus: "The vision of a cosmos developing from the lowest types and stages upward through beast and man to higher and better man is enchanting and almost overwhelming. This worldview had to me at first a charm and witchery that was almost intoxicating. It created more of a revolution than an evolution in my thinking. But more careful consideration convinced me that the little truth in it served to sugar-coat and give plausibility to some deadly errors that lurked within."

W. H. Griffeth Thomas, Wycliffe College, Toronto, Canada, at present contributor to the Sunday School Times, has this to say: "As to a theory of evolution regarded as a process of uninterrupted differentiation of existences, under purely natural laws, and without any Divine intervention, it would suffice to say that it is 'not proven' in the sphere of natural science, while in the realms of history and literature it is palpably false."

Now from all these considerations we must recognize the fact that evolutionists do not constitute a homogenous party; and that while there is a prevailing tendency to acknowledge the reality of organic evolution as extended to the species and production of all living forms, there is likewise a growing disposition to question the sufficiency of causes by which Darwin sought to account for it. And there is likewise a growing tendency among scientists and other scholars to deny or abandon the whole theory and to accept the primitive Divine creation theory in which it is held with Buffon the French naturalist (1707–1788) that an original progenitor of certain great groups of organisms "participated in the grace of a distinct act of creation," from which, through variation and adaptation developed the multitudinous forms of animal and plant life. For instance, all wolves and dogs may be allowed to be descended from a pair of ancestral wolves; all of the plants called "liverworts" and their close affinities from an ancestral form; all human beings from an ancestral pair. In this case, the original progenitor of each great group was totally unconnected with every other.

I may venture at this point to remind you, though probably the caution is needless, that the question as to whether the present popular theory of evolution is or is not a fact, is not purely a scientific one as a number of scientists would have us believe, but it is fundamentally a religious one to be judged by the Word of God. For God's revelation to us in the Bible must correspond with His handiwork in nature. The mind of man is so organized that whenever he seeks evidence for some favorite belief, no matter how erroneous, he is able to find it. What God has spoken to us in His Divine, inspired and authoritative Word is our only criterion here as elsewhere. To repeat what I have written elsewhere, "The Scriptures preclude any other doctrine than that of an absolute creation of all things by the direct act of the Divine will. Before this truth, all the errors of pantheism, polytheism, dualism, materialism, atheism, and spiritualism fall." (Bible Doctrines, page 48.)

Lecture II—Studies in Science

In the last lecture, the question at issue was placed squarely before us. In this it is our purpose to examine certain claims made by evolutionists, the establishment of which is necessary to the successful maintenance of their scheme. As was suggested, the question of evolution is primarily a Biblical one; yet, apart from the Bible and on the principles of true science, it can be shown that the present popular theory is fallacious. This is what I shall proceed to accomplish.

I. THE ORIGIN OF LIFE

The essence of modern evolutionary theories is that man, brutes, and plants, i. e., all living things, originated from one simplest and lowest form of organic life, by purely natural and godless selections and variations. To again quote Prof. Packard in the New International Encyclopedia: "Life appears to have been a necessary and inevitable result of inorganic or cosmic evolution. It came into being on our planet in the most natural way.... It arose by the action of physicochemical laws through what we call spontaneous generation, the materials for the first bit of protoplasm being ready at hand." By "ready at hand," professor Packard undoubtedly means existing in the elements of inorganic materials. "This places us at once face to face with the question of ...".

Spontaneous Generation

By spontaneous generation is meant life coming from lifeless material in a purely natural way. For instance, it was at one time supposed by some scientists that the millions of living forms, which may be obtained from placing a handful of hay in water and letting the whole stand for a few days, came about from the decomposition of the hay. That life came from the lifeless. Now, of course, we know that instead of decomposition causing life, living organisms cause decomposition. Where, then, do these organisms come from? To answer this question, we must refer briefly to the history of the spontaneous generation theory.

In 1675, Anton van Leeuwenhoek, the Dutch microscopist, discovered, to use his own quaint language, "living creatures in rainwater, which had stood for four days in a new earthen pot, glazed blew within." He was an ardent follower of Harvey, and was one of the first to offer experimental evidence against the then-current belief that the lower organisms arise spontaneously, and to establish the truth of Harvey's dictum, "omne vivum ex vivo," all life is from life. He later discovered bacteria, and at once the question arose, Do these generate spontaneously or are they descendants of beings of the same kind? In 1749 Needham claimed that he had by experiment proved that bacteria developed spontaneously. But Spallanzani, Italian anatomist, in 1769, called attention to the careless methods employed by Needham. Schultz, Dusch, Schwann, Schroeder, Hoffman, Pasteur and Tyndall all proved by open experiment that the presence of living organisms is due always to the pre-existence of similar living forms. Dr. Bastian, an English biologist, in 1871, again secured a very decided support for the origin of living matter de nova or from the vitalization of non-living matter. But shortly afterward Cohn of Breslau demonstrated that some bacteria had the ability to go into a resting or spore stage in which they are able to resist drought and other deleterious influences. This was the death-blow of spontaneous generation, and we now know that nothing is more certain than the truth of Harvey's law, "omne vivum ex vivo." In the words of Dr. E. B. Wilson of Columbia University: "There is no evidence that 'spontaneous generation' or the production of living from lifeless matter without the influence of antecedent life ever takes place." (General Biology, page 33.)

Since this theory of spontaneous generation has been exploded, other terms have been invented to take its place. Such terms as "bathmism," "germ-plasm," "cosmic ether," "pangenesis," "cosmic emotion," "vital fluid," "growth force," and many others used about twenty years ago, suggest the strenuous efforts to account for the origin of life from purely natural means. They are no longer seriously employed by any reputable scientist. They have all disappeared, as has Huxley's (1825–1895) word for the process, "abiogenesis." Prof. Haeckel of Jena is the most ardent and open supporter of this exploded theory, and many scientists are still searching for proof of the theory which seems to die hard. Once in a great while we hear a statement such as issued a few years ago from Prof. Loeb of Chicago who said: "The Divine creation of life is a pure humbug. Life originally happened. I believe that in a short time real life will be created in the laboratory." But then those who believe this are ashamed to say much about it and the rest soon forget about it. Among hundreds of others who deny the theory we may mention Agassiz, Carpenter, Beale, Dana, Dawson, Faraday, Gray, Herschel, Helmholtz, Lord Kelvin, See, Leibnitz, Latze, Pasteur, Verdt, Haeke, Maury, Romanes. These ascribe to God alone the power to originate life.

Life Depends on Life

The theory of evolution not only presupposes the unproven theory of spontaneous generation but denies that life depends on life. In other words, it assumes that life in its lower forms could subsist without the higher forms. Suppose we were to grant the theory that life originally came by natural processes from the inorganic world, how would it have maintained its existence? This is a point that has always appeared to be the greatest barrier, outside the Bible, to a belief in evolution, and yet, strange to say; I find practically nothing written about it.

To make this point clear notice how everything living depends on everything else living. Plants grow from the ground. Animals feeding on plants form food other animals, and the so-called lower forms of life, especially unicellular organisms, have a definite relationship in the sustenance of life with both the animal and vegetable kingdoms. This interdependent relationship is absolutely necessary for the existence and continuation of all living forms. How then can the evolutionist say that life began with a simple cell when that simply by means of nature could not exist by itself?

But in order to convince my scientific friends on this important matter, I must offer real scientific proof. "The bacteria have no chlorophyll and are therefore unable to obtain nitrogen and carbon from such simple bodies as carbon dioxide and ammonia; they must get their carbon and their nitrogen, as such, from organic matter. The bacteria are as active in their life cycle as the so-called higher organisms. They cause profound changes in the organic matter on which they live. Developing in the living body, the parasitic bacteria cause disease and death. The saprophytic bacteria cause decomposition, putrefaction, and fermentation, and thus they resolve the tissues of dead organisms into the simpler compounds, carbonic acid, water, and ammonia, in which form they serve as nutrition for the more highly organized members of the vegetable world, without which simpler compounds these plants cannot live. They cannot get their carbon and nitrogen from the complicated substances which are the foodstuff for bacteria The bacteria furnish the green plants with the simple compounds which these can use, the plants that possess chlorophyll having the power in the presence of sunlight to break up these compounds into their constituents. Were it not for the activity of the bacteria, which furnishes food to the plants, there could be no plant life, and as all animal life is dependent directly or indirectly on the vegetable world, there could be no life at all upon earth. The so-called 'low' and the so-called 'high' forms of life are interdependent; they necessitate and presuppose one another. Without the higher forms of life, no lower forms are possible, and without the lower forms there can be no higher forms. Every living being depends on every other living being." (Schultz) How then can there be a single low form or even a few, which evolutionists insist upon? There cannot.

The Cell is Complex

This leads up to the fact that the cell is not simple but very complex, which offers another and still harder blow to the Darwin-Haeckel-almost-everybody-else theory. For this type of evolutionist insists that the primordial cell was simple and that evolution proceeds from the simple to the complex. Now this cell, which they say was simple must at the same time have been able to do what it takes unicellular organisms, plants, and animals all working together to perform. How could this supposed living thing exist by itself? Would a marvelous thing like this be simple? Certainly not. The truth is that in proportion as this thing was simple it is the more absurd to think of it as being able to live alone. And if it was such a marvelous thing that could do of itself all that it takes all living things in conjunction to do, then it was not simple, but the most complex, the most elaborate, the most compound, the most highly organized affair that ever existed in fact or in imagination.

So when the evolutionist tells us that organic progress has been from a simple cell to more complex, he needs to be instructed that the living cell is as complex as any other organism and that the more the biology of the cell is being investigated the more complex the living cell is shown to be.

Now the unicellular organisms which are not considered so highly organized by evolutionists are still more complex, for we find cilia or flagella by which they move and in many of what are known popularly as the lowest forms a most bewildering and complex life history. What must have been the "simplest" cell from which in the mind of the evolutionists all these living forms came?

Possibly it will be thought that I have spent too much time on the question of the origin of life by natural means as is so fallaciously held. But I consider this fundamental, and if the foundation of any system is shaken, the whole superstructure is bound to fall. We have seen that spontaneous generation exists nowhere. As E. D. Cope has pointed out, this theory is a nonsensical conception. He says: "It is more probable that death is a consequence of life, rather than that the living is the product of the non-living."

II. THE EVOLUTION OF SPECIES

Definition of Species

If we could grant evolutionists their idea of the origin of life, other claims which they erroneously make would cause the system to fall, and foremost among these is the theory of descent as it is held with reference to the origin of the species. What do we mean by species? De Candolle, the eminent botanist, says: "We unite under the designation of a species all those individuals that mutually bear to each other so close a resemblance as to allow of our supposing that they may have proceeded originally from a single being (in the case of the lower organisms) or a single pair." There is a great difference of opinion among scientists as to what constitutes a species, but this is pretty generally accepted by naturalists. The idea is more tersely expressed by Buffon, who says: "A species is a constant succession of individuals similar to and capable of reproducing each other." Possibly the most convenient definition for us in this consideration is that of Cuvier, the great French zoologist. He describes a species as: "A succession of individuals which reproduces and perpetuates itself." With this definition it is very easy to divide the higher animals in a general way into different species. Dogs and wolves belong to the same species because they interbreed. Horses and hippopotami do not interbreed and thus belong to different species even though "hippus" means a horse.

No Transmutation of Species

Now it is rather strange that, in the face of this definition which precludes transmutation of species, the evolutionist insists on the belief that by natural processes one species of plant, or animal, may be transformed into another and that by long-continued and progressive changes the higher types of life, culminating in man, have been evolved from the lower. And this is still more strange when we remember that there is not a single case on record anywhere of such a change. Yet this is a fact which nearly every writer on descent admits. Prof. Conn says: "It is true enough that naturalists have been unable to find a single unquestioned instance of a new species. It will be admitted at the outset on all sides, that no unquestioned example has been observed of one species being derived from another." (Evolution of Today, page 23). Prof. Winchell writes on this point as follows: "The great stubborn fact which every form of the theory encounters at the very outset is, that notwithstanding variations we are ignorant of a single instance of the derivation of one species from another. The world has been ransacked for an example, and occasionally it has seemed for a time as if an instance had been found of the origination of a genuine species by so-called natural agencies, but we only give utterance to the admissions of all the recent advocates of derivation theories, when we announce that the long-sought experimentum crucis has not been discovered." (The Doctrine of Evolution, page 54.)

Crossing the Species

Just as there is no case on record of a single instance of the evolution of a new species, so there is no knowledge of the successful crossing of species. Among the twenty-some thousand species of animals classified not one case is known where different species have been crossed which did not result in sterility of the animal thus begotten. As Dr. Townsend says: "If this always has been the case, and no reason can be given for thinking otherwise, then there is shut out completely what seems to be the most available agency at nature's command for the production of new species."

The fact is very well known that animals of different species do not naturally breed together. When under domestication man succeeds in overcoming their repugnance for each other the offspring are sterile among themselves, although in some instances they will breed back to the pureblood. All the modification of any species thus produced, or by any means, can be brought about and invariably happens in a short time; and thereafter these variations revert to the normal type.

The importance of this point is so great to the subject in hand that I wish to quote from Dr. Robert Patterson, who says: "So universally admitted is the fact of the sterility of hybrids of different species, that the evolutionists have labored diligently to find some case of such sterility between different varieties produced from the same species by selective breeding. Since, according to the theory, varieties are only incipient species, this should be quite practicable and even common. But no such case can be found." (Errors of Evolution, page 226.)

Variation of Species

The idea contained in the quotation just given, that varieties are only the beginnings of new species, is also unproven. To be sure such men as Hugo de Vries, of the University of Amsterdam, and Luther Burbank in California, appear to have developed mutable species. The former has been working with primroses and the latter with roses, berries, and plums. Prof. Standfus of Zurich, by variations of food and temperature, is claimed to have obtained several new species of butterflies. These changes and others such as are constantly being made at our universities with pigeons, dogs, and other animals and plants, are admittedly not new species. Many of the experimenters do not regard them as such. But the radical evolutionists seize at anything they can find to substantiate their unproven theory. As direct proof of this let me quote from the great botanist Hugo de Vries, just mentioned, in connection with primroses. He says: "Fluctuating variation cannot overstep the limits of the species even after the most prolonged selection—still less can it lead to the production of new permanent characters." M. Blanchard declares in his work as accurately as we can translate it from the French: "All experimentation and observation point out that, while the mutability of organisms in a state of nature shows itself in various degrees, yet, in its most wonderful manifestations, it remains confined within a circle beyond which it is impossible to pass." (La vie des etres animes, page 102.)

The Mutation Theory and Heredity

De Vries was led by his experiments in plant heredity to the conclusion that changes might take place much more suddenly than had been supposed. This is known as the mutation theory and such a departure from type, developed suddenly in a single generation, de Vries spoke of as a "mutation." Many evolutionists thus believe that species have arisen in mutations, that is in sudden variations—offspring differing in some marked and well-established characteristics, for example, short-horn cattle, Ancon, and Mauchamp sheep, turnspit dogs, Japan peacocks, primroses, etc. Now many leading biologists, among them some of the rankest evolutionists, do not admit that mutations help the various theories of evolution. The great T. H. Morgan, Columbia University, says, "The mutation theory destroys species; it does not originate them." (Experimental Zoology, page 233.)

On the whole subject of variation whether gradual or sudden, we must admit that there is much change, but that this variability has proved quite illusive under close examination. It is observable where many different varieties of the same species are found together, crossing freely. A close study has shown that large differences cannot accumulate from the accumulation of small differences. But what is still more to the point is that a close study of heredity has shown us that the differences between these varieties are "factorial"—that is, that the various individuals possess in varying proportions certain constituents of the original breed from which all have diverged. The salient point is whether it can be shown that small variations accumulate in such a way as to cause transformations from one form to another across the dividing lines between species, genera and larger divisions. Professor Bateson said in his Presidential Address in 1914, before the British Association: "We have done with the notion that Darwin came latterly to favor, that large differences can arise from the accumulation of small differences. This is quite definite and certainly authoritative enough. But he continues: "Such small differences are often mere ephemeral effects of conditions of life, and as such are not transmissible; but small differences, even when truly genetic, are factorial like the larger ones, and there is not the smallest reason for supposing that they are capable of summation." Indeed the results produced by artificial and experimental breeding are all factorial, and what is still more to the point the differences are not brought about by the addition of new factors but by the loss of factors and an extreme emphasis of one factor, for example the large udder of the cow and the excessive egg-laying of the hen. Let a few illustrations clear up this matter. First, listen to what H. T. Edge has to say on moths (Sept. 1916, Theos. Path): "Somewhere there exists a parent moth from which all other moths sprang, and whose germinal cells contain all the factors which have since become separated and distributed in varying proportions in the germinal cells of the descendants. Perhaps that parent animal no longer exists. In either case, the evidence from the study of heredity points to the conclusion that the differentiation is rather on the downward grade than the upward." Again, this same authority tells us: "Instead of all the domestic fowls being improvements on the old jungle-fowl they are merely shattered fragments of that efficient old bird. Thus we have now in our barnyards fowls of the most fantastic complexity, but they are by no means improvements on the original parents. They are mere factors, simulacrums, hopeless digressions. All they can do in the way of breeding is to go on producing more fowls and by some sudden understandable event an occasional mutation, opening up still further vistas of the latent possibilities in that original hen."

So, then variations whether slight or marked, whether arising gradually or suddenly, whether occurring naturally or artificially, cannot be shown to cause transformations from one form to another passing over the dividing line between species, genera and other divisions.

Argument from Geology

The facts mentioned thus far are enough to upset the whole theory, but since evolutionists persist in producing evidence pointing, as they think, to the evolution of the species, we must notice some which they consider the strongest, our first consideration being the testimony of geology. Geology is the history of the earth and its inhabitants, as read in the structure of the earth's crust. Since, as we have seen, the evolution theory claims that life came upon the earth by purely natural causes from the inorganic matter, and then developed through long-continued processes into all the numerous forms living and extinct, it is but proper that we should turn to the record of life as we gather it from the fossil remains from the earth's crust. Does the evidence of the geological record—which, as Huxley said, is the only direct evidence on the question of evolution—does this evidence stand for or against the origin of existing species from extinct ones by means of minute gradual changes, which we must remember is the only vital consideration?

The argument of historical geology is that the fossilized remains of former life appear in the strata of the earth in advancing order, the simplest followed by the more complex. The assumption is that the higher grew out of, the lower, by a chain of infinitesimal changes long continued.

Time will not permit a long discussion here, but in a brief review, I wish to say that the geological record of life on this globe contradicts evolution. There has not been the infinite time which the theory of evolution demands. The order in which living forms have appeared according to the geological record shows that new forms appear abruptly instead of gradually; that the higher orders often appear before the lower and the larger before, the smaller. There are at least four fatal gaps in the gradation of the species which are impassable; first, between dead and living matter; second, between plants and animals; third, between different species; and fourth, between unreasoning brutes and rational human beings. No theory of evolution can bridge these chasms. Darwin, Huxley, Haeckel, and others have undertaken it, but the timbers of their bridges are all too short.

By a careful survey of geological evidences of life, scientists of late are discovering a mass of evidence pointing in the exact opposite direction of universal progress, so that scarcely a word is heard on that subject. Such words as retrogradation, deterioration, degeneration, and reversion are becoming more common than universal progress. And it is found that by a careful application of the principles of evolution to geology, the verdict of the best geologists is that evolution has been shattered on the geologic rocks.

The Missing Link

Evolutionists have been much concerned about these missing links which they need to fill out the great gap in the geological record. Darwin, Huxley, Haeckel, and others in their writings berate the imperfections of the record. But today there is a tendency on the part of evolutionists to keep quiet on this subject because they see that it is too imperfect by half to answer their purposes. As Patterson says: "Let us note that for every pair of distinct geologic species in our cabinets, we should have about 20,000 intermediate forms; and that these missing links are conspicuous by their absence." (Errors of Evolution, page 225). Dr. Winchell tells us: "The forms intermediate between them (family types of geology) are universally wanting. One cannot understand why in all rocks whatever, and in all countries upon the two continents, all the relics of the intermediate types should have vanished. The absence of intermediate forms characterizes the gaps between genera and even species as well as between orders and families." (The Doctrine of Evolution, page 1404.) The fact of the matter is, as Patterson says, they never had any existence save in the brains of some enthusiasts. They are only ghosts seen in the mind's eye. But as they are vital to the evolutionist's theory with their disappearance his theory melts into such stuff as dreams are made of.

Pithecanthropus Erectus

As is natural, there is no richer field for the evolutionist on the missing link problem, than that of finding the link connecting man with the brutes, the most common relic being pithecanthropus erectus. Dr. Von Eugene DuBois, a Dutch physician, in 1891, found a piece of skull, two teeth, and a leg bone, at widely separated places, and at different times, on the island of Java. This is all there is to this wonderful connecting link. A few years after the discovery, twenty-four of the most eminent scientists of Europe met at Leyden and passed their verdict on these bones. Ten said they were nothing but the bones of a monkey, seven that they were those of a man, and seven pronounced them really the missing link between man and the ape. Thus only seven, not one-third of the twenty-four, ascribed any importance whatever to the relic.

The Colorado Specimen

From Prof. Stephen Bowers, of the Geological Survey of California, we have the following solution of a skeleton found in Colorado: 'A few years ago the newspapers contained an account of the discovery of a skeleton in Colorado by a Columbia College professor, which he was pleased to call the missing link between man and the apes. He gave this remarkable creature an antiquity of a million and a half of years. The friable bones were carefully wrapped in cotton and shipped east, but scarcely had the learned professor got away with his prize when certain cowboys came forward and claimed the bones to be that of a pet monkey which they had buried but a dozen years previously.

Skulls

Two skulls deserve a passing notice: the Calaveras skull in the California State Museum, and the Neanderthal skull found in Prussia. Concerning the first R. C. Scribner, a shop keeper, acknowledged to W. H. Dyer, a clergyman of Los Angeles, that he placed the skull in a deep mine shaft as a practical joke. Dr. Meyer of Bonn examined the evidence concerning the Neanderthal skull, claimed to have been from 200,000 to 300,000 years old, and found it to be a Cossack killed in 1814.

Thus, it will be seen; the theorists are endeavoring to solve a very great problem on the strength of very meager evidence. The few and scattered remains mentioned cannot be accepted as sufficient to support any theory whatever as to human evolution. The various theorists are at odds with one another over these "evidences." So far as it is possible to obtain a general idea from these remains, it sums itself up in the conclusion that, throughout the whole extent of time over which these remains extend, the earth was peopled with human beings, not differing on the whole from those which inhabit it today, and that man remains true to his single species.

Bones and other remains are today regarded with much suspicion. In criminal cases, when the recent remains of human beings are examined experts are often unable to decide whether they are human or brute. With T. H. Morgan, our foremost American zoologist, we can say about all this matter of missing, links: "There is no proof that the common plan of structure must be the result of blood relationship."

Likenesses

This likeness in structure just referred to is another point held by evolutionists to support their theory. The argument is that low adult forms and embryos of higher forms are strikingly alike. An embryo of a reptile passes through the changes of a fish, and the human in the germ cannot be distinguished from any other mammal. Here the evolutionist gets excited, jumps up from his microscope and shouts aloud to all the world that all organisms, even plants, are blood relations. But his glass has failed him. The apparently similar cells are very different. For out of one comes a mouse, out of another a monkey, out of another a whale, and out of another a man, and nature never blunders in the matter. We often see pictures of the development of the locomotive with Stevenson's "puffing billy" first, and then through all grades up to the greatest locomotive in the world. Who would say that the one grew from the other just because there are some resemblances? Two pictures hang side by side on the wall. Was the smaller evolved from, the larger? No! The same artist painted both. So, in nature it is the warm touch of the Maker in His universal style.

Rudiments

From the examinations of these similarities has arisen the theory of rudimentary organs. There are organs, the evolutionists say, which in our supposed animal existence were of use to our ancestors but are now useless. For instance, the hairy covering of our supposed patriarchal ancestors is now no longer needed. The evolutionist rolls up his sleeves and notices that the scanty covering of hair all turns back toward the elbow. He remembers that that is the way it is in the monkey. Then it dawns upon him that the monkey is his ancestor and that the hair took that position because his great-great-great-great-great-grandfather sat in the branches of the trees and the rain kept washing the hair back that way until it kept its place. There seems to be plenty of evidence here for the evolutionist, for Dr. Wiedersheim in one of his evolutionistic fits has found 107 rudimentary organs in man. Houston Chamberlain says it is about time to write a book on the human body as nature's junk shop for defunct organs and wonders whether the human brain is not the 108th rudimentary organ. As these organs are declared useless, they are cut out for trivial reasons. The thyroid gland is a rudimentary organ, "cut it out" and out it comes. But many of the victims died from the effects of the operation; all the others developed myxedema, a form of cretinism. It was rudimentary, was it? In concluding this point, let me quote Dr. A. P. Schultz, Monticello, N. Y. "The Darwinian theory," he says, "muddled the brain of anatomists, physiologists, and surgeons, it confused the unsuspecting public. The great sale of Haeckel's works speaks well neither for Haeckel nor for the public. The theory of evolution is a consequence, a symptom, and a cause of the materialism which is rampant; a materialism as crude and as superficial as that which infested imperial Rome." (The End of Darwinism, page 15).

Natural Selection and Adaptations

I have only time to glance briefly at natural selection and adaptation. Suffice to say that it cannot accomplish those variations which occur in opposition to the physical forces. It cannot produce new organs. The simpler forms still live and have not been crowded out of being by the superior types. Variations are not minute and gradual but often great and sudden. Accidental variations could not produce the minute and ingenious adaptations of nature as seen in the eye. Plan and purpose must have gone before variation. The struggle for existence is not a (source of elevation but issues only in degradation. Hunger does not make men nobler but brutalizes them. Famine develops Cannibalism, and it is to be noted, that in this respect at least man is below the anthropoid ape for they are never found to be cannibalistic. Take an earthworm and cut it in two in the middle, the posterior piece grows at its cut end, not a head but a tail. Can it get along better with two tails than one? No; its environment has led to its deterioration. Exactly the opposite is true in certain flatworms. If the head of a certain planarian is cut off just behind the eyes, there develops at the cut surface, not a tail as we should think, but a new head, facing in the opposite direction. Thus, the struggle for life does not account for adaptive structures. Prof. T. H. Morgan, referring to these experiments performed in his class at Columbia says: "These and other reasons indicate that regeneration cannot be explained by the theory of natural selection."

Degeneration

From this, it is seen that progress requires something more than adaptation to the environment. Darwin himself admitted that natural selections could not account either for origins or for progress. Evolution may explain the survival of the fittest, but not the arrival at the fittest. The fact of the matter is that we see not so much evolution upward as an evolution downward. We may say a devolution. This was at one time a good word. It is now considered archaic. I think it should be resurrected for this occasion. In all nature there is no law of universal improvement. All things have their beginnings and endings. The four d's, dawn, development, decline and death, state one of nature's most exacting laws, which is no less true of the vast universe or of the physical part of man, than of the insect that lives but a day. Multitudes of species, plants, and animals show no development when compared with their earliest types. Man from a biological viewpoint shows no improvement. Permanence, degeneration, improvement followed by reversions is the order of the day. I shall have more to say on this subject in our next lecture.

Persistence

I have just used the word permanence, and this deserves a word of explanation and makes a fitting conclusion to this whole matter. You have noticed that the evolutionist insists that change for the better is the order of the day. I have already intimated that this is not true, but that persistence in spite of tendencies to degenerate seems to be the most vital consideration. Let me explain briefly. In the opening of this lecture, I made clear that every living being is the product of another living being. It is Harvey's dictum "omne vivum ex vivo," all life from life. Now I want to go a step further and say every living being is the product of a living being of the same kind. The decree of Almighty God away back, in the beginning, has not changed one particle. He said, "After its kind." That stands. We all know that this is true with the higher order of animals and plants with which we are familiar. Let me tell you it is also true of the lower orders. Never mind the understanding of these words but get the thought anyway. You know enough to get the idea. Listen! A chromogenic bacterium comes from a chromogenic bacterium; a photogenic bacterium comes from a photogenic bacterium. You, scientific agriculturists if you are here, know that a nitrifying bacterium comes from a nitrifying bacterium. You, doctors of medicine, know that diphtheria producing bacterium comes from diphtheria producing bacterium, and not from tuberculosis producing bacterium, and so on. Were this not so we could have no science of bacteriology. If the Divine fiat "after its kind' would no longer hold what would we have? Confusion and total extinction. Do you see what this means? Catch the force of it for it is final and convincing. This means that persistence of the species is the essence of life and not change of species as the evolutionist holds.

Lecture III—The Bible and Evolution

In this lecture, we wish to show that just in proportion as the evolutionary idea of descent is unscientific it is unbiblical. If evolution were truly scientific it would be biblical; since it is unscientific it is also unbiblical.

I am well aware of the fact that some scientists and even some theologians will object to making this subject a biblical one. But if evolution maintains theories contrary to the Word of God, this question certainly becomes biblical. It is eminently a biblical question since radical evolutionists and atheists have hitherto lifted it up triumphantly as a scientific disproof of the very foundations of the Christian faith. But we have nothing to fear. True science and the Bible will agree. Since God is the Author of both Nature and Scriptures, these two must agree. They do agree. Man may not always be able to fully harmonize the account of Scripture with apparent facts in Nature, but this is simply because of human limitations. Nevertheless, the fact remains that what is written in Nature agrees with what is written in God's book—the Bible. Primarily the Bible is not a treatise on science. It is very much higher than science. Galileo, the noted astronomer, truthfully says: "The Scriptures were given not to tell us how the heavens go, but to teach us how to go to heaven." Yet the facts found in the Bible are thoroughly scientific. A certain writer of this age says: "Science has had to acknowledge, over and over again, the correctness of the creation account, and withdraw the objections and assaults which have been made." (A. C. Gaebelein). And without a doubt science will ever continue on this course, because she is young and very fallible, whereas God is eternal and infallible.

I wish to develop the thought of true science in contrast with that which is false. There are three great principles in the world which concern the scientist;

viz., matter, energy, and life. The first two give rise to the physical, and the third to the biological sciences. In physical science, the first and fundamental law is the conservation of matter and energy. This means that matter and energy may be transformed, but can never, except by miracle, be created nor destroyed. This is biblical and sound reasoning. But when we come to biological science, where we deal with life in connection with matter and energy, we are confronted with false science in the form of un-teleological evolution, which says that life is either matter or energy or a combination of matter and energy. This is not biblical and therefore untrue and unscientific. For the very first words of the Bible are: "In the beginning God created the heaven and the earth." We thus have the creation of living things. Life is thus neither matter, nor energy, nor a combination of these, but a third and distinct principle. Just as science substantiates conservation of matter and conservation of energy, so it substantiates conservation of life. But certain scientists fail to regard life as a distinct created principle, as the Bible teaches, and so we have "science falsely so-called" in connection with biological work, viz., un-teleological evolution.

Origin of Life

Now we are at once placed face to face with the first point under consideration—the contradiction of evolution and the Bible as to the origin of life. As we have seen, radical evolutionists attempt to explain the origin of life from purely natural causes. I repeat that the question of how life originated is similar to the question of how matter and energy originated. Scientists have never pretended to know more about this matter than what is revealed in God's Word, "In the beginning God created." Why should they attempt to know more about life? For the evolutionist to insist in attempting to demonstrate how life originated is as senseless as for scientists to try to demonstrate how matter became inert or energy changeable. The very thought is unreasonable and absurd. The German author, Goethe, whom Matthew Arnold styled "our greatest modern man," said: "Lebhafte Frage nach der Ursache ist von grosser Schaedlichkeit"—"Anxious questioning as to the cause is pernicious." In this case, we agree with Bacon when he said, "Causas non oportet investigare"—"We ought not investigate causes." The idea in these two quotations is that no amount of investigation will land us anywhere. All we can know is "God created." With this we should be content.

The Origin of Man

Nor is evolution's account of the origin of man in accord with teachings, but it is purely heathen teaching. This may be shown by comparing side by side the Bible, evolution, and a typical heathen account.

Bible—Gen. 1:26, 27; 2:7; 5:1, 2.

"And God said, let us make man in our image, after our likeness; and let them have dominion over the fish of the sea, and over the fowl of the air, and over the cattle, and over all the earth, and over every creeping thing that creepeth upon the earth. So God created man in his own image, in the image of God created he him; male and female created he them. And the Lord God formed man of the dust of the ground and breathed into his nostrils the breath of life, and man became a living soul. In the day that God created man, in the likeness of God made he him; male and female created he them; and blessed them, and called their name Adam, in the day when they were created."

Darwin's Descent of Man—Chap. 2, page 372.

"Man is descended from a hairy quadruped, furnished with a tail and pointed ears, probably arboreal in its habits and an inhabitant of the Old World. This creature, if its whole structure had been examined by a naturalist, would have been classed among the Quadrumana, as surely as would the common and still more ancient progenitor of the Old and New World monkeys. The Quadramana and all the higher mammals are probably derived from an ancient marsupial animal, and this through a long line of diversified forms, either from some reptile-like, or some amphibian-like creature, and this again from some fish-like animal. In the dim obscurity of the past, we can see that the early progenitor of the Vertebrata must have been an aquatic animal, provided with branchia, with the two sexes united in the same individual."

Religions of Ancient China, Giles—pages 7, 8.

"Before the beginning of all things, there was nothing. In the lapse of ages, Nothing coalesced into Unity, the great Monad. After more ages, the great Monad separated into duality, the male and female principles in Nature; and then, by process of biogenesis, the visible universe was produced. Male and female principles were each subdivided into Greater and Lester: and then, from the interaction of these agencies, a being named Pan Ku came into existence. His breath became the wind; his voice the thunder; his left eye the sun; his right eye the moon; his blood flowed in rivers; his hair grew into trees and plants; his flesh became the soil; his sweat descended as rain; while the parasites which infested his, body was the origin of the human race."

As we have said repeatedly, the whole theory hinges on this idea of the descent of man from the brute. The Bible denies this; the evolutionist and the heathen believe it. The Bible brings God in as the Creator of man; the evolutionist and the heathen leave Him out, the latter because he has no knowledge and the former because of unbelief. The Bible brings man into existence fully developed; the evolutionist and the heathen evolve him from organic connections through diversified forms.

Some may say, the Bible record is regarded by some as much of a myth as the heathen idea. True, some hold this idea, and none stronger than the evolutionist. If this is true the whole Bible collapses and we are utterly undone. For then, the testimony of Jesus Christ Himself is false. Christ refers to no less than a dozen books and some twenty narratives of the Bible to the subjects of Creation (Matt. 19:4); Law of Marriage (Matt. 19:5); Cain and Abel (Matt. 23:35); The Flood (Matt. 24:37); Abraham (Jno. 8:56); Lot's Wife and the Destruction of Sodom and Gomorrah (Luke 17:28–32); Manna (Jno. 6:49); Brazen Serpent (Jno. 3:14); Elijah and his Miracles (Luke 4:25, 26); Tyre and Sidon (Matt 11:22); Jonah and the Whale (Matt. 12:39); The Books of Moses (Jno. 5:46); The Psalms (Luke 20:42); Moses and the Prophets (Luke 24:27); and the entire Old Testament (Luke 24:44). Not once does Christ refer to these events as myths but as facts. He gives them a literal interpretation. Whom are we going to believe? The evolutionist, the heathen with not a single proven fact to their credit, or the Son of God, "The Way, the Truth, and the Light," the Word which "liveth and abideth forever?"

Savagery—Degeneration

Furthermore, this evolutionistic view of the origin of man asks us to believe that man came from the brute to a race of cannibals, and up through savagery to fraternalism and philanthropy, which again is unbiblical. The Bible account of man's first state represents him as a child, but never as a savage. The essentials of man's original state are summed up in the Bible in the phrase, "the image of God." This implies his natural likeness to God or personality; and moral likeness to God, or holiness. This surely is very different from a primordial cell or even a brute, which evolutionists say, "rubbed off its tail by repeatedly sitting down and which could stand on its hind legs and throw things with its forelegs." According to the Bible man when created was lord of nature. He named the animals. He had a moral sense which could be appealed to. He delighted in occasional communion with his Maker. He had right intuitions and inclinations and was entirely free to choose between good and evil.

In spite of this biblical and therefore proper view, evolutionists insist upon two things which they say facts corroborate, that man's prehistoric condition clearly shows a progress from primitive savagery to civilization; and that the religious history of mankind points to a continuous progress—a universal law—whereby man passes from fetichism to polytheism and monotheism.

With reference to the first view, it is necessary to call attention to the fact that it is based upon insufficient evidence. History, biblical and secular, plainly shows a law of degeneration running by the side of development and often even counteracting it. Workers in the field of research along anthropological lines have come to agree with Prof. Schlegel, who was one of the first to advocate degeneration, and who said: "The more I investigate ancient history, the more I am convinced that the nations set out from a true worship of the Supreme Being." We may go back as far as it is possible in history and we find nations in a very high state of civilization. In Gen. 4:16–22, we read of a very ancient civilization. Dr. Scofield, in his notes, comments upon this passage thus; "The first civilization, that which perished in the judgment of the flood, was Cainitic in origin, character, and destiny. Every element of material civilization is mentioned, city and pastoral life, and the development of arts and manufactures. Enoch, after whom the first city was named, means 'teacher.' The el termination of the names of Enoch's son and grandson shows that for a time the knowledge of Elohim (gods) was preserved, but this soon disappeared (Rom. 1:21–23). Adah means 'pleasure.' or 'adornment;' Zillah. 'to hide;' Lamech, 'conqueror,' or 'wild man.' The Cainites civilization may have been as splendid as that of Greece or Rome, but the Divine judgment is according to the moral state, not the material." This is shown by the Scripture found in Gen. 6:5, 7.

So we see as Dr. Strong says: "In the earliest record, we find nations in a high state of civilization; but in the case of every nation whose history runs back of the Christian era—as for example, the Romans, the Greeks, the Egyptians—the subsequent progress has been downward, and no nation is known to have recovered from barbarism except as the result of influence from without." (Systematic Theology, page 270).

Here then, history again verifies facts as given in the Bible. See Rom. 1:21–32, concerning the Gentile nations.

This then is the scriptural and true scientific view. Not a growth from brute to cannibal and on to civilization, but vice versa, from a civilized state down to cannibal, and finally to a brute-like condition. It is interesting to note a few more scientific facts which substantiate this view. Gulick tells us that, "Cannibalism and infanticide are unknown among the anthropoid apes. These must be the result of degradation. Pirates and slaveholders are not men of low and abortive intelligence, but men of education, who deliberately throw off all restraint, and who use their powers for the destruction of society." Mark Hopkins says: "There is no cruel treatment of females among animals."

This same process of deterioration noticed now in the organic and social realms, holds in the consideration of the movement of religious thought. You may take any religious system you choose and the further back you trace it the purer it is from polytheistic elements. The students who took the mission study work last year on comparative religions as set forth in Dr. Speer's "The Light of the World" will appreciate the truth of this statement. Thus, all evidence points to an original monotheistic basis for all religions. Our missionaries tell us that if the heathen were left to themselves, they would show no signs of growing better, but of continually becoming worse. I sum up this point by quoting Dr. Tisdall in his work "Christianity and Other Faiths" (page 211): "Wherever we are able to trace any religion back for many ages, we find it purer, nobler, and loftier in its teachings about God in the earliest times than in the latest. This is true of Hinduism, the early religion of China, the primitive faith of the dynastic Egyptians, and in many other instances."

So then, in all these considerations, we see that there is no universal law of improvement, organically, socially or religiously, such as the evolutionists claim, but the evidence is on the side of the biblical record of degeneration. Outside of Christ, evolution thus becomes devolution, and from the reading in Romans, we noticed even devolution. Tennyson, who, though an evolutionist, was not a dangerous one when he said:

"Evolution, ever climbing after some ideal good,
And reversion, ever dragging Evolution in the mud."

Jesus Christ

I just mentioned Christ, and under this heading, I believe I can say all that is yet to be said. If the evolutionists regard Christ at all, they must somewhere along the line of descent make due provision for Him. In the long period which has elapsed since our Lord's life on earth, evolution ought to have produced still more perfect characters, nobler teachers, holier and better men than He; or at least it should have made the type normal. It will hardly be contended that this has been so. As more than nineteen hundred years have passed since Christ's birth, and as, even with His example to imitate, humanity has never yet produced any other character, in the least even to be brought into comparison with Him, there must be something radically wrong in the theory that would evolve life from the molecule of atom, and pass it through all the processes of growth and development, through the physical up to and including the spiritual. Here is the greatest gap in the evolutionist's theory. Over and above the fact that no animals use tools, make fires, and talk stands the fact that man is a religious animal. Where along the line of descent does the evolutionist place the incoming of a religious nature? By no manner of natural means can this be accounted for. So, we come back to the good old Book—God's Word. "In the image of God created him." The evolutionist tries to rule God out of the beginning of his theory of progress. His cry is "purely natural causes." He seeks to rule Him out of the middle. In fact, neither beginning, middle, nor end shall be supernatural. Or he may admit Him a little at one point in order to give color to his system. Or he may in a very liberal way admit a sprinkling of Him all along the line for the sake of policy. Invariably and inevitably, the evolution theory of descent leads to this atheistic or pantheistic beginning from which it sprung. Invariably and inevitably, the personal God and the Christ of God, as we said of Auguste Compt's philosophies, are conducted to the front for a respectable introduction and then are bowed out of their universe with thanks for their provisional services. As Dr. Strong says: "The grievous error of this modern overstatement of evolutionism is that it divorces the phenomenal from the noumenal, makes bodily change a rule for the soul, makes science as vain as the cat's pursuit of its own tail, turns the universe into a medley of accidents without law and without God."

Conclusion

It is hardly necessary, to sum up, points. But in a brief statement I gather up this whole matter. Evolution theories of descent are all fallacious; they are unsupported by scholars; they are admittedly not proven; they are not scientific; they are unbiblical, and they are either atheistic or pantheistic.

I feel perfectly free to accept all that true science can prove with regard to the origin and development of man; for this development must be a theistic development, with Christ as its agent and goal. So that while I claim a Christian evolution, an evolution in which Christ is the center and circumference, yet, this evolution is as far different from the purely automatic and necessary development, as is day different from night. I claim for the word evolution a Christian use, but this use is in no way compatible with the idea which the modern evolutionist holds of continual change and the denial of permanence.

Now I wish to make an appeal. Most of us are students; many are teachers or preparing to teach; some of us are parents. This subject is of the greatest importance to all of us. The teaching of evolutionistic theories as approved science, to the children from the public schools to postgraduates of the universities all over the world, is the most lamentable feature of this whole matter. The Darwinians hesitated to launch the idea of descent. They were fearful of the consequence. Virchow tried heroically to protect the schools of Germany. In his great speech at Munich, he plead, "We ought not to represent a conjecture as a certainty, nor a hypothesis as a doctrine." But the radicals, foremost among them Haeckel, the greatest living advocate of organic descent, advocated it, and as Henry Beech says in Vol. VIII of "The Fundamentals:" "As they struggled someone lighted the fire. It was like the burning of the temple of Jerusalem. Titus had issued an order to spare it, but a Roman soldier, threw a blazing torch into a small window and the whole structure was in flame." The late A. R. Wallace, who advanced the theory of descent at the same time as did Darwin, said shortly before his death, in his ninety-first year: "I have come to the general conclusion that there has been no advance either in intellect or morals since the days of the earliest Egyptians. Everything is as bad as it possibly can be. There exist in our midst horrors and dreadful diseases never known before. Our whole social environment is rotten, full of vice and everything that is bad." As we have seen, others have taken the same view. Let us keep the good work moving. We look for a still greater reaction, although as Dr. Wright has said recently, "Doubtless the passing of the present phase of evolution is not final."

It is well that this series of lectures should conclude with the Bible. It is well that the preacher's parting word should be of faith in the revealed Word of God, which giveth light. Hold it in its wholeness as the best teaching on all subjects which the world has yet seen or ever will see; true on all subjects, even though partial on some, and full enough on every subject necessary to educate us for time and eternity. Hold the Bible, then, the most sacred and best, not only for its certainty—surer than any other knowledge—not only for its deep and wide philosophy of men, of things animate and inanimate, of the universe; but far away and above all other reasons hold it as the revelation of a Savior for the lost and degraded world.

Made in United States
North Haven, CT
31 December 2025

86098154R00060